GENETIC ENGINEERING

GENETIC ENGINEERING

ANIL KUMAR
AND
NEHA GARG

Nova Biomedical Books
New York

For permission to use material from this book please contact us:
Telephone 631-231-7269; Fax 631-231-8175
Web Site: http://www.novapublishers.com

Library of Congress Cataloging-in-Publication Data

Genetic Engineering / Anil Kumar and Neha Garg. (Editors).
p.; cm.
Includes bibliographical references and index.
ISBN 1-59454-753-X (hardcover)
1 Genetic Engineering
WL 360 P9643 2005] I. Anil Kumar and Neha Garg
RC377.P762 2005
614.5'9834--dc22 2005002153

Published by Nova Science Publishers, Inc. ✣New York

Contents

Preface

The present book has the distinguishing feature of having condensed material with adequate information on genetic engineering especially of microbes. The book covers almost all the topics of genetic engineering for the graduate, postgraduate student and young research scholars of biological sciences. The book is written as a syllabus of genetic engineering for a Masters course in biotechnology, biochemistry, life sciences at most of the universities. The book is most useful for Masters degree students. Emphasis is given on the basic fundamentals. The book contains twelve chapters starting wtih 'Isolation, purification and estimation of nucleic acids' in chapter 1. The chapter describes general techniques for the isolation and purification of DNA as well as RNA. It also describes methods for quantitative estimation of the nucleic acids. The second chapter describes general characteristics of the vectors used in genetic engineering and also the general account of commonly used individual vectors. The chapter also describes expression vectors. The third chapter describes various commonly used restriction endonucleases. The fourth chapter describes commonly used enzymes in genetic engineering viz. Reverse transcriptase, DNA polymerase I, polynucleotide kinase, teminal dcoxynucleotidyl transferase, alkaline phosphatase, SI nuclease, DNA ligase etc. The fifth chapter describes electrophoresis for the separation of nucleic acids fragments. The sixth chapter is of cloning strategies. It describes construction of a genomic DNA library , chromosomal walking, cDNA library, cDNA cloning. The seventh chapter describes DNA sequencing techniques and includes the chemical modification method of Maxam and Gilbert, dideoxy sequencing method of Sanger, modifications of chain terminator sequencing, analysis of the sequencing data. The eighth chapter includes various methods of site directed mutagenesis. The ninth chapter describes polymerase chain reaction (PCR). It also includes primer designing and various types of polymerase chain reactions viz. reverse transcriptase polymerase chain reaction (RT-PCR), nested PCR, multiplex PCR etc. Also, there are chapters 10, 11 and 12 on gene therapy, human genome and proteomics. At the end, is a glossary which explains the main terms used in genetic engineering. One of the important factors introduced in the book is the chapter structure given in the beginning of each chapter that provides, at a glance, the contents of the whole chapter which offers a better learning mechanism. Each chapter is also presented with an introduction that covers the concept of the whole chapter in brief and offers a clear understanding of the subject matter to the students. The author (Anil Kumar) on the basis of his experience in teaching genetic engineering at the university level for more than a decade has offered the text in an easily understandable form to the postgraduate student. The book should be of invaluable help to the students, researchers and all those interested in understanding genetic engineering.

Chapter I

Isolation, Purification and Quantitation of Nucleic Acids

1.1. Introduction

Isolation of nucleic acids is considered an important protocol and use of any vigorous technique for rupturing of the cells should be avoided. It is due to the reason that mostly for many researches, intact nucleic acid molecules are to be extracted without shearing. After rupturing the cells, further care should be taken for isolation of the nucleic acids without shearing. Purity of the preparation is also important since impure preparations are not generally preferred for various protocols of genetic engineering (like constructing genomic library, cDNA library, site directed mutagenesis, recombinant DNA technology). One should also be familiar that in molecular biology and genetic engineering, all the glasswares, plastiwares, reagents etc are sterilized before use. While working, the person should always wear gloves.

1.2. Isolation of Nucleic Acids from Prokaryotes (Microbes)

Isolation of Chromosomal DNA from the Bacteria

The bacterial cells are grown at the optimum temperature in a suitable medium depending upon the bacteria and its strain. The cells are harvested by centrifugation at 10,000 xg for 10-15 minutes in the cold condition, washed with a suitable medium in order to remove any adhering traces of the growth medium and care should be taken that cells should not rupture. For *E. coli*, the most studied bacteria, generally150 mM NaCl containing 15 mM sodium citrate buffer, pH 7.0 (buffer A) is used. Following centrifugation, the cell pellet is suspended in the same medium. Generally, *E. coli* isolated from 500 ml growth medium is suspended in 15 ml of the buffer A. The suspended cells are treated with lysozyme for conversion into spheroplasts. Lysozyme ruptures the cell membrane by digesting the glycosidic bonds. For the purpose, the suspended cells are incubated with lysozyme (nearly 1 mg) for few minutes. Afterwards, the cells are lysed using sodium dodecyl sulfate (SDS) (generally 0.2 ml of 25 % SDS). Thereafter, 200 μg/ ml RNase (which has already been heat treated at 80°C for 10 minutes to inactivate DNase activity if any, otherwise DNase will digest DNA) is added to digest RNAs. The RNase is prepared as 10 mg/ml solution in 0.1 M sodium acetate containing 0.3 mM EDTA, pH 4.8, and heated at 80°C for 10 minutes. Thereafter, pronase or proteinase K enzyme is added to chew the proteins. Generally, 10 mg proteinase enzyme is added and the preparation is incubated at 37°C for few minutes. Following incubation, to increase the pH of the system, addition of 1M tris-HCl, pH 9.0 to a final concentration of 50 mM is preferred. Thereafter, equal volume of phenol- chloroform – isoamylalcohol in the ratio of 100:96:04 is added and mixed thoroughly by vortexing and incubated in the cold condition for about 20 minutes. During this period, proteins get denatured. The sample is centrifuged to separate the aqueous layer containing DNA and organic (phenol) layer. Proteins get collected at the interface. The upper aqueous layer containing DNA is removed carefully without disturbing the interface and the lower layer, and is transferred into an another container. Phenol treatment is repeated twice. Thereafter, equal volume of water saturated diethyl ether is added to remove traces of phenol, if any. Diethyl ether being lighter forms upper layer and can be removed carefully without removing the aqueous layer. Treatment with diethyl ether is repeated once. High quality molecular biology liquid phenol is used in the process and is treated by the following procedure called as Sevag method of phenol treatment before use.

Sevag Method of Phenol Treatment

To the liquid phenol, equal volume of 1M Tris is added and shaken thoroughly. The upper aqueous layer is removed by aspiration and the process is repeated till the pH of the phenol is neutral or slightly alkaline. Thereafter, equal volume of TEN buffer (10 mM tris, 1 mM EDTA, 100 mM NaCl, pH 8.0) is added, shaken and left at room temperature to separate the layers. Aqueous layer is separated by aspiration and the step is repeated once. Afterwards, phenol is treated with TE (10 mM tris, 1 mM EDTA, pH 8.3) buffer and phenol is stored in cold condition over TE buffer. It is preferable to add a pinch of 8-hydroxyquinoline (antioxidant) in the phenol preparation. It stabilizes the phenol against air oxidation. However, some researchers do not prefer to add antioxidant since 8-hydroxyquinoline is yellowish in color and it becomes difficult to distinguish whether pinkish color is due to antioxidant or due to oxidation of phenol to quinone which also gives pinkish tint in phenol. If phenol develops pinkish tint, then phenol is discarded.

After phenol treatment, DNA is collected at the end of a glass rod after addition of two volumes of ethanol. Chromosomal DNA being in the higher concentration gets stick on the glass rod. Addition of sodium chloride to a final concentration of 0.2 M or sodium acetate, pH 4.8-5.2, to a final concentration of 0.25 M is preferred followed by two volumes of ethanol and stored in cold condition at –90°C (dry ice-ethanol bath) for 20-30 minutes. As a result of this procedure, there is precipitation of DNA which may be collected by centrifugation in the cold condition. DNA pellet is washed with 95 % ethanol once and dried under vacuum. DNA may be dissolved in TE buffer.

Isolation of Plasmid DNA (Miniprep)

The following protocol is used for miniprep that yields approximately 10 μg of plasmid DNA from 1 ml culture. The plasmid DNA is 80-90% pure and may be used for restriction endonuclease digestion studies.

1. One ml of the stationary phase culture of the transformed bacteria (*E. coli*) is transferred in a clean 1.5 ml eppendorf tube and cells are pelleted by centrifugation at 11,000 x *g* for 5 minutes at room temperature using a microfuge. The supernatant is discarded and the pellet is washed with 1 ml of SET buffer (20% sucrose, 50 mM EDTA, 50 mM tris, pH 7.6). Afterwards, the pellet is suspended in 150 μl of SET buffer.
2. 5 μl of RNase (10 mg/ml, prepared as described above) is added followed by 350 μl of the lysis buffer (0.2 N NaOH and 1 % SDS) and mixed gently. The contents are kept in an ice bath for few minutes till there is precipitation of SDS which is indicative of white precipitate.
3. 250 μl of 3 M sodium acetate, pH 4.8 is added, mixed and incubated on ice for 30 minutes. At this stage, there is precipitation of chromosomal DNA which is evident from the thick viscous precipitate. Here, different conformational stages of plasmid and chromosomal DNAs discriminate plasmid DNA that is not precipitated out by sodium acetate but it precipitates chromosomal DNA.
4. The contents are centrifuged at 11,000 x *g* for 10 minutes at 0-4 °C in a microfuge to separate chromosomal DNA precipitate. Plasmid DNA is in the soluble form. The supernatant is removed carefully in a separate clean eppendorf tube. Care must be taken that not even a tiny particle of the precipitate should be in the supernatant after transferring into another eppendorf tube.

700 μl of the supernatant can be transferred safely from step 4 in another eppendorf tube. An equal volume of isopropanol is added and after mixing by upside down, it is stored for 5-10 minutes at the room temperature. At this stage, there is precipitation of plasmid DNA by isopropanol. The amount of the plasmid DNA is too small to make it invisible by naked eye. The contents are centrifuged at 11,000 x *g* for 10 minutes at the room temperature. The invisible pellet is washed with 80% ethanol and dried under vacuum in a Speed Vac. The dry pellet is dissolved in 20 μl of water or TE buffer. Two to 5 μl of this solution may be used for restriction endonuclease digestion.

Now a days, many companies sell kits having ready made reagents. Using the kits, one can isolate plasmid DNA in lesser time. In the kits, mini spin column and collection tubes are also used. The mini-spin column have controlled porous membrane. These spin column are used to retain plasmid and thereafter the spin-column membrane having plasmid is washed with wash reagent (having mainly alcohol). Afterwards, the plasmid DNA is dissolved either in TE buffer or simply in nuclease free water. Since, we do not precipitate plasmid DNA using alcohol, it saves time. However, the basic principle with the kits is also the same.

Large Scale Preparation of Plasmid DNA

The basic principle of isolation of the plasmid DNA in this protocol too is same as in miniprep. However, here further purification of the plasmid DNA is done to get pure plasmid DNA for cloning/ sequencing purpose. The same protocol is for both bacterial plasmids (without cloned fragment) and recombinant plasmid. The following steps are involved in the protocol for isolation from one litre of tha culture:

1. The culture is harvested by centrifugation at 10,000 x*g* for 10 minutes in the cold condition. For harvesting one litre culture, four 300 ml capacity autoclaved bottles may be used, each will be having approximately 250 ml culture. After harvesting, the supernatant is discarded and the pellet in each bottle is washed with 30-50 ml SET buffer. After washing, the pellet in each bottle is suspended in 30 ml of SET buffer. Now the following steps may be carried out in each bottle unless mentioned otherwise.
2. Lysis buffer (70 ml) is added in each bottle and incubated at 50°C for 30 minutes with occasional swirling in a water bath. During incubation, there is rupturing of the cells. After 50°C incubation, the bottle is incubated in an ice bath for 30-40 minutes and during this incubation, excess SDS gets precipitated out.
3. To each tube, 50 ml of 3 M sodium acetate, pH 4.8 is added and incubated in the ice bath for about 40-50 minutes. During this incubation, chromosomal DNA gets precipitated out whereas plasmid DNA remains in the soluble form.
4. The contents are centrifuged in the cold condition at 10,000 x*g* for 10-15 minutes to settle the precipitated chromosomal DNA and the supernatant is poured into another bottle after filtering through Kimvipes. The purpose of filtration through Kimvipes is that not even a single tiny particle should go with the supernatant (use of Kimvipes is optional).
5. To the supernatant, 0.2 ml of RNase (10 mg/ml) is added and incubated at 37°C for 15 minutes. During this incubation, RNA of the bacteria is cleaved by the RNase.
6. 50 ml of phenol-chloroform-isoamylalcohol is added for extraction of proteins which get denatured and get collected at the interface. The contents are centrifuged in the cold condition at 10,000 x*g* for 10 minutes in order to separate the layers completely. The upper aqueous layer is carefully separated with the help of a pipette and is transferred into an another clean bottle. The contents of the aqueous layer containing plasmid DNA are kept at the room temperature.
7. An equal volume of isopropanol is added to the aqueous contents and stored at the room temperature for 10-15 minutes to precipitate plasmid DNA. The plasmid DNA due to its low quantity, is not visible. In fact, if DNA is visible in the bottle, it indicates the possibility of contamination of the plasmid DNA with the chromosomal DNA. The contents are centrifuged at 10,000 x*g* for 15 minutes at the room temperature to pellet the plasmid DNA. The supernatant is poured off carefully without disturbing the invisible pellet of the plasmid DNA. The minute amount of the liquid is removed by putting the bottle in the inverted condition over the paper towel / Kimvipe. Afterwards, traces of the liquid are evaporated by keeping the

bottle in a vacuum oven at 55°C for 2-3 hours. At this stage, the bottle should be completely dried.

8. The pellet is suspended in 8 ml of TE buffer. The solution of all the four bottles may be poured in one bottle. Then, the total volume in one bottle will be 32 ml. To it, 42 g of cesium chloride (molecular biology optical grade) is added and dissolved by incubating at 37°C for few minutes with slow shaking.
9. Refractive index of the solution is measured in a refractometer. It should be between 1.390 and 1.395. If it is more, the solution is diluted by adding TE buffer dropwise. If the refractive index is lesser, more cesium chloride should be added.
10. To the solution, 1.1 ml of ethidium bromide (5 mg/ml) is added. Ethidium bromide intercalates between the bases of the DNA. After addition of ethidium bromide, the solution should not be exposed to light and afterwards, it is put in an ultracentrifuge for centrifugation at 42,000 rpm for 24 hours at 20°C.
11. After centrifugation, the tubes are carefully removed from the rotor of the ultracentrifuge without disturbing the bands of the nucleic acids. In a dark room, the tube is kept in vertical condition and ultraviolet (UV) light is thrown on the tube with the help of an UV torch. Upon exposure of the UV light, the whitish bands are visible in the tube. In the lower half, there will be RNAs. In the upper half, generally two bands are visible near to each other. The lower band is of closed circular double stranded plasmid DNA which is removed very carefully by puncturing the tube with the help of a needle connected with a syringe. Care should be taken that contents from the upper band should not get mixed with the contents. The contents after taking into the syringe are transferred into an another tube. To the contents, equal volume of n-butanol is added and mixed thoroughly. Butanol extracts ethidium bromide from the solution. Butanol forms upper layer which may be removed with the help of a pipette. The process is repeated till all the ethidium bromide gets removed from the DNA solution which is indicative from the color. The solution will be colorless when no ethidium bromide is there.
12. The aqueous solution containing plasmid DNA is dialyzed in the cold room against 1000 fold excess of TE buffer overnight with at least one or two changes. The dialysis removes traces of ethidium bromide, if any and also cesium chloride. After dialysis, the contents are transferred into an another tube and 5 M NaCl is added to a final concentration of 0.2 M followed by two to three volumes of 95% chilled ethanol and stored at –70 to –90°C (dry ice ethanol bath) for 30 minutes. If stored in a deep freezer, then it should be stored for several hours preferably overnight for complete precipitation of the plasmid DNA. Afterwards, the contents are centrifuged in the cold condition to settle the precipitate of plasmid DNA. The pellet is washed with 95% ethanol for removing any adhering impurity. Thereafter, the contents are dried in an vacuum oven at the room temperature for evaporation of traces of ethanol. The dried pellet is dissolved in one ml of the TE buffer. For quantitation of DNA, 5 to 10 μl sample is diluted to one ml and absorbance measurement is made at 260 and 280 nm. The ratio of 260 to 280 nm should be between 1.8 to 2.0 if plasmid DNA does not have contamination of proteins.

Nowadays, people do not prefer to do large scale purification of plasmid DNA. Many researchers prefer to perform PCR for amplification. The PCR technique is easy and quick compared to large scale purification of plasmid DNA. It also does not require ultracentrifuge. After PCR amplification, the PCR amplicon is purified. Nowadays, PCR product purification kits are available commercially. Some researchers prefer to run PCR product on agarose gel electrophoresis. Afterwards, band of PCR amplicon is cut and DNA is extracted for which, gel extraction kits are available.

1.3. Separation of DNA and Rnas

If a sample contains DNA and RNAs (tRNA, rRNA, mRNA), the various components may be separated by density gradient centrifugation either using cesium chloride gradient or sucrose density gradient.

With cesium chloride, density gradient is formed during centrifugation. The molecules having different buoyant density are separated as separate bands. Buoyant density of RNA is much higher (1.9 g/ml) than single stranded DNA (1.71 g/ml) and double stranded DNA (1.70 g/ml). RNA gets settled down in the lower half of the tube. Single and double stranded DNAs having different buoyant densities also get separated from each other. As described above for the large scale plasmid isolation, plasmid DNA can be separated from the chromosomal DNA on ethidium bromide-cesium chloride density gradient centrifugation. Since circular plasmid DNA binds lesser ethidium bromide, therefore has higher buoyant density than linear chromosomal DNA which binds more ethidium bromide and has lesser buoyant density.

Sucrose density gradient centrifugation is carried out to separate different types of RNAs. Different RNAs differ in their size, therefore these have different sedimentation rates. During centrifugation, each type of RNA gets separated as a layer which may be removed by puncturing the tube at the bottom and collecting fractions of fixed volume (in the form of droplets). The absorbance of individual fraction may be read in the spectrophotometer and different peaks of different RNAs may be marked. Here, a gradient of 5 to 40 % sucrose in TE buffer is prepared in the polypropylene tube and RNA sample is layered over it. The centrifugation is done at 80,000 x *g* for about 20 hours in the cold condition to separate various types of RNAs. The fractions constituting a single peak for a particular type of RNA are separated and concentrated by ethanol precipitation as described above.

Separation of different types of nucleic acids can also be achieved by using either ion exchange, gel filtration or adsorption chromatography. A common ion exchanger used is methylated albumin kieselguhr (MAK). Commonly used gel filtration media are Sephadex. A common adsorption media is hydroxyapetite. Now a days, in high performance liquid chromatography systems/ fast protein liquid chromatography systems, suppliers supply specific columns for separation of nucleic acids.

Polyacrylamide / agarose gel electrophoresis is also used for the separation of DNAs and/or RNAs of different sizes. Sometimes, polyacrylamide gel electrophoresis under denaturing conditions is used to separate the two complementary strands of a double stranded DNA. If the nucleic acids are of less than one kilobase in size, polyacrylamide gel electrophoresis is used while for separation of nucleic acids of more than one kilobases, agarose gel electrophoresis is used. Nowadays, agarose gel electrophoresis is preferred over polycarylamide gel electrophoresis even for smaller size DNA fragments. Since nowadays, gel extraction kits are available commercially and these kits have solutions having capability of dissloving agarose gel easily. After separation of the nucleic acids on the electrophoresis gel, the bands of the nucleic acids are visualized after staining with ethidium bromide which binds with the nucleic acid molecule. After staining with ethidium bromide, the nucleic acid bands show white fluorescence upon exposure to ultraviolet light. The bands containing a particular nucleic acid may be cut from the gel with the help of a sharp blade and elution of the nucleic acid may be carried out by electro-elution or using gel extraction kit. Some researchers prefer to add ethidium bromide in the agarose gel itself (during preparation). In electroelution, the gel containing nucleic acid is put inside a dialysis bag along with one to two milliliters of tris-acetate buffer (0.04 M tris-acetate containing 0.001 M EDTA; 50x buffer is prepared by dissolving 242 g of tris base, 57.1 ml of acetic acid and 100 ml of 0.5 M EDTA, pH adjusted to pH 8.0 and volume adjusted to one litre). The dialysis tubing after sealing is put in a submerged type of electrophoresis chamber and is submerged with the same tris-acetate buffer. A current of nearly 100 mA is passed through the buffer. As a result, the nucleic acid moves from the gel and comes in the buffer inside the dialysis tubing. After one hour or so, one can check with the help of ultraviolet light in a dark room whether the gel inside the dialysis tubing still has nucleic acid which will be indicated by fluorescence. If still nucleic acid is present in the gel, electroelution is carried out for some more time. The buffer containing nucleic acid is collected in a microtube and phenol-chloroform treatment is done to remove impurities of proteins/ gel traces and precipitation of the nucleic acid is done by ethanol precipitation as described above. Nowadays, electro-elution is not a preferred technique due to low yield and more time consumption. One can isolate DNA from the gel band using commercially available gel extraction kits. One gets more yield and time consumed is lesser with the kits. The kits are of two types. In some kits, spin column with collection tube is used whereas in others, beads are used for binding DNA.

1.4. Methods for Estimation of Nucleic Acid

(A)DNA

(I) Phosphorus Method

DNA is a polynucleotide molecule made of deoxy nucleotides. These deoxyucleotides are consisted of purine/pyrimidine bases, deoxyribose and phosphate molecule. The phosphorus is in the organic or stable form. Therefore, DNA may be estimated by quantitative phosphorus estimation methods. The DNA is first digested with concentrated sulfuric acid over a digestion rack for 6 to 8 hours to convert it into inorganic phosphorus. For the purpose, DNA along with sulfuric acid in Kjeldahl flask is digested on a digestion rack. The care is taken that at no time, drying of the contents inside the flask should be there which is maintained by the addition of sulfuric acid. When DNA is digested over the digestion rack, after about 6 hours, there is blackening inside the flask indicative of the presence of carbon. The contents are heated more till all the carbon gets completely oxidized to carbon dioxide (there is disappearance of the black material). After digestion, the contents are transferred quantitatively in a volumetric flask and volume is made up with the help of distilled water. The phosphorus is estimated in this solution (unknown sample) by any of the well known methods of inorganic phosphorus determination. Although less sensitive, still the most common method of phosphorus determination is Fiske Subbarow's method.

Fiske Subbarow Method for Inorganic Phosphorus Estimation

In this method, the basic principle involved is conversion of phosphorus into phosphomolybdic acid in the presence of sulfuric acid and ammonium molybdate. Phosphomolybdic acid gets quantitatively converted into molybdenum blue (various oxides of molybdenum) in the presence of 1-amino, 2-naphthol, 4-sulfonic acid (reducing reagent). During this reaction, phosphorus acts as a quantitative catalyst. That's why, color intensity increases with time on storing the tubes. The color intensity is measured using UV-VIS spectrophotometer.

The following reagents are required for the estimation:

1. 10 N Sulfuric acid
2. 2.5 % ammonium molybdate, aqueous solution. To increase its stability, few drops of sulfuric acid are added in it.
3. Reducing agent: To 195 ml of 15 % sodium metabisulfite or 13.7 % sodium bisulfite, o.5 g of 1-amino,2-naphthol,4-sulfonic acid is added and dissolved by lukewarm heating. On cooling, if still there is any insoluble material, the same may be removed either by decantation or by filtration. To it, 5 ml of 20% sodium sulfite is added and mixed thoroughly. The solution is stored in a brown bottle to avoid exposure of the light. The solution is stable upto a month or so.

In the aliquot (unknown sample), water is added to make it 0.84 ml. To it, 40 μl of 10 N sulfuric acid is added and mixed thoroughly. Afterwards, 80 μl of 2.5 % ammonium molybdate is added and mixed. In the last, 40 μl of the reducing agent is added and mixed. The contents are incubated for 10 minutes at the room temperature and then color intensity is read in a spectrophotometer at 660 nm. A standard series of phosphorus is also run simultaneously under the identical conditions. By comparison, one can calculate the amount of phosphorus in the unknown sample.

Nowadays, more sensitive methods are also available for phosphorus estimation. In these methods, mostly stannous chloride is used as reducing agent.

Method Based on Sugar Estimation (Deoxyribose Present in DNA)

(II) Diphenylamine Method

The method is used for determination of the amount of deoxyribose in DNA which will be direct indicative of the amount of DNA. The following reagents are used in this method:

1. **Diphenylamine reagent:** one g of purified diphenylamine is dissolved in acetic acid and volume made to 100 ml with the acetic acid. Afterwards, 2.75 ml of concentrated sulfuric acid is added for stabilization.
2. **Standard DNA solution:** One mg/ml DNA solution is prepared in water. Afterwards, equal volume of 5 % perchloric acid is added and heated at 70°C for 15 minutes, cooled and the volume is made up to get 0.5 mg/ml DNA in the sample. Perchloric acid treatment is done to hydrolyze DNA. Unknown DNA sample is also treated in the same way.

Up to 400 μl of the aliquot may be taken (if the aliquot is less than 400 μl then volume of the aliquot is made to 400 μl with water). Afterwards, 800 μl of diphenylamine reagent is added and the tube is placed in a boiling water bath for 10 minutes, cooled and the color intensity is read in a spectrophotometer at 660 nm. A series of tubes having standard DNA concentrations is also run simultaneously and by comparison, one can calculate the amount of DNA in the unknown sample.

(III). Spectrophotometric Method

It is considered to be the practical method for estimation of DNA in genetic engineering experiments since here DNA can be estimated without destroying the sample whereas in the other two methods mentioned above (phosphorus method and diphenylamine method), sample is destroyed during estimation. Here, absorbance of the sample is measured in a spectrophotometer at 260 nm. For high molecular weight DNA, an absorbance of 1.0 is taken equivalent to 50 μg DNA/ ml sample (if the light path in the cuvette is one centimeter). With low molecular weight oligonucleotides, the value is taken 35 μg per ml sample. The purity of the DNA sample is checked by the ratio of absorbance at 260 nm and 280 nm which should be between 1.8 to 2.0.

Some investigators determine the amount of the DNA by using the following formula called as Kalchar formula:

DNA (mg/ml) = 1.45 OD at 260 nm - 0.74 OD at 280 nm

(B) RNA Estimation

(I). Phosphorus Method

The method is used as described above for DNA. However, RNA's phosphorus is alkali labile. It can be converted into inorganic phosphorus by heating at 100°C for 15 minutes in the presence of 1N NaOH. After hydrolysis, the sample may be neutralized by adding sulfuric acid before estimation.

Method Based on Sugar Estimation (Ribose Present in RNA)

(II). Orcinol Method

In RNA, there is ribosyl moiety. Ribose is measured by orcinol method which will be indicative of the RNA concentration. As in case of diphenylamine method, here too a standard series of RNA is run for comparison. The orcinol method requires the following reagents:

1. **Orcinol reagent:** 6 % orcinol in 95 % ethanol.
2. **Acid reagent:** To 100 ml of concentrated HCl, 0.5 ml of 10 % $FeCl_3.10H_2O$ is added.
3. **Standard RNA solution:** To an aqueous RNA solution (1 mg/ml), equal amount of 1 N NaOH is added and heated in a boiling water bath for 15 minutes, cooled and neutralized using HCl and the volume is made to get working solution of 50 μg RNA /ml.

To the aliquot, 200 μl in volume (RNA solution plus water), 200 μl of orcinol reagent and 600 μl acid reagent is added and after mixing the tube is heated in a boiling water bath for 20 minutes, cooled and the color intensity is measured at 665 nm. The concentration of RNA in the unknown solution may be calculated by comparison with the standard RNA.

(III). Spectrophotometric Method

It is same as given for DNA except here it is considered that 40 μg/ ml RNA will give an absorbance of 1.0 in a light path of one centimeter.

1.5. Purity of Nucleic Acid Sample

As described above, in genetic engineering without wasting the sample of the nucleic acid, purity of the sample is determined by comparing the absorbance at 260 nm and 280 nm. The ratio of 260 to 280 nm absorbance should be between 1.8 to 2.0.

1.6. Isolation of Nucleic Acids from Eukaryotes

Protocol for Isolation of DNA from Blood

1. Blood (10 ml) is collected in the tubes coated with anticoagulant so that blood should not clot. Generally EDTA is used as anticoagulant. However, citrate-dextrose (0.48 g citric acid, 1.32 g sodium citrate, 1.47 g dextrose, water to 100 ml; 1 ml of this solution is used for 6 ml blood) is a preferred anticoagulant. The blood may be used immediately for isolation of DNA or it may be stored at 0°C for several days or at –90°C indefinitely before extraction of DNA. Blood is centrifuged at about

1500x*g* for 10-15 minutes to collect the blood cells as pellet and supernatant (plasma) is discarded. A buffy coat present in the tube is transferred to an another tube and recentrifugation is done. This buffy coat contains white blood corpuscles of heterogenous density. Buffy coat after second centrifugation is suspended in 15 ml of the buffer containing 10 mM tris, 10 mM EDTA, 20 μg/ml RNase, and 0.5 % sodium dodecyl sulfate (SDS), pH 8.0. The suspension is incubated at 37°C for one hour to lyse the cells.

2. Proteinase K is added to a final concentration of 100 μg/ml and is mixed thoroughly but gently.
3. The contents are incubated at 50°C for 3 hours in a water bath. This incubation is done to get the proteins released and hydrolyzed by proteinase K.
4. The contents are cooled to room temperature and phenol chloroform isoamylalcohol treatment is done in the same way as described above for isolation of DNA from microbes to separate the nucleic acid from proteins. The denatured proteins get collected at the interface of the phenol (lower) and aqueous phase and DNA is in the aqueous phase.
5. For isolation of larger size DNA (more than 200 Kb in size), the aqueous phase after phenol extraction is dialyzed against 50 mM tris, pH 8.0 containing 10 mM EDTA until the OD of the dialysate is less than 0.5. Absorbance at 260 and 280 nm is also measured to check the contamination of proteins. If no contamination is there, ratio of the absorbance at 260 and 280 nm should be between 1.8 to 2.0. If from the absorbance, it looks that protein contamination is there, SDS to a final concentration 0.5% is added and again all the steps from step 2 onwards are repeated. If low molecular weight DNA has to be isolated, DNA is precipitated out by the addition of ethanol as described above for extraction of DNA from microbes. Normally, from 10 ml blood, there is a yield of about 125 μg of DNA. The procedure can also be used for isolation of DNA from cultured mammalian cells. On an average, from 5 x 10^7 cells, there is yield of about 200 μg of DNA.

Protocol for Isolation of DNA from Plant Leaf Tissue

1. Leaf tissue (5 g) is washed and is put in either liquid nitrogen for maceration to a fine powder. Some researchers prefer to add a pinch of glass beads during grinding. The powdered sample is immediately transferred in 25 ml of the extraction buffer (Urea 42 g; 5 M NaCl 7.0 ml; 1 M tris, pH 8.0, 5.0 ml; 0.25 M EDTA 8.0 ml; 10 % sarkosyl, 20 ml phenol (prepared after treatment as given above in Sevag method) , and water to make the volume 100 ml) pre-incubated at 65°C. The contents are mixed in the extraction buffer and stirring is continued so that no clump formation occurs. Afterwards, 0.75 ml of 20 % SDS is mixed and incubated at 65°C for 10 minutes.
2. To the contents, 15 ml chloroform-isoamylalcohol (24:1) is added and mixed thoroughly by inverting the tube upside down and is incubated at room temperature for 10-15 minutes.

3. The contents are centrifuged at 5000 x*g* to separate the layers. The upper aqueous layer is removed and transferred into an another tube. An equal volume of isopropanol is added and mixed by inverting the tube upside down. Nucleic acids get precipitated out with in few minutes and are collected on a glass rod by mixing the contents with the glass rod. Afterwards, nucleic acid collected, is dissolved in TE buffer. If it does not dissolve at the room temperature, it may be incubated at 60°C up to one hour to dissolve it in TE buffer. The preparation may be having contamination of RNAs.
4. To the preparation, RNase (final concentration 10 μg/ml) is added and incubated at 37°C for about 30 minutes to digest RNA, if any. Afterwards, phenol chloroform treatment is done as described above for removing contaminant proteins. Thereafter, DNA is precipitated out by adding sodium acetate (0.25 M final concentration) and two volumes of 95% chilled ethanol as described above.

Isolated DNA is stored in a -20°C freezer. For longer storage, it is preferred to store in a –70 to -90°C freezer.

Protocol for the Isolation of RNA from Yeast

After harvesting, yeast cells are suspended in distilled water and incubated at 37°C for 15 minutes. If one starts with about 25-30 g of yeast cells, these may be suspended in 100-120 ml of water. Afterwards, TE saturated phenol (160 ml) is added and stirred for 30 minutes at the room temperature. Thereafter, it is centrifuged for 10 minutes at 5000 x*g* in the cold condition to separate the organic (phenol) and aqueous layers. In yeast, DNA remains associated with the proteins in the form of nucleoproteins. Upon phenol treatment, there is dissociation of nucleoproteins. DNA remains in the lower phenol phase whereas RNA is in the upper aqueous phase. Denatured proteins remain at the interface.

RNA containing aqueous phase is removed carefully and potassium acetate is added to a final concentration of 20 g per litre. Afterwards, using ethanol, RNA is precipitated out in the same way as precipitation of DNA described above. Generally, RNA constitutes about 4% of the dry weight of the yeast cells.

1.7. Isolation of Total Cellular RNA and Fractionation of mRNA from Eukaryotes

The Following Protocol May Be Used for Isolation of RNA from Plant Tissue

The basic general principle in isolation of RNA from plant tissues is same as in yeast. DNAs remain as nucleoproteins and upon phenol treatment, dissociation of nucleoproteins occur. The DNA remains in the phenol phase and RNA in the aqueous phase which may be removed and RNA may be precipitated out using ethanol.

Plant tissue is ground with the extraction medium in a pestle and mortar. Generally, 10% homogenate is prepared using the extraction medium (1% SDS, 0.15 M sodium acetate, 5 mM EDTA, 50 mM tris-HCl, pH 9.0). To the homogenate, an equal volume of phenol-chloroform-isoamylalcohol (100:96:04) is added for dissociation of proteins. The contents are thoroughly mixed and incubated for 5 minutes at the room temperature. Afterwards, it is centrifuged at 10,000 x*g* for 10 minutes to separate the layers. The upper aqueous layer containing RNA is removed carefully and ethanol precipitation of RNA is done in the same way as described above. Special precautions should be taken to avoid contamination of RNase. For the purpose, all the glasswares, plastiwares should be rinsed with 1% diethylpyrocarbonate to inactivate RNase.

Protocol for the Fractionation of mRNA from Total Cellular RNA

After isolation of total RNA as described above, to isolate mRNA, advantage of the presence of polyA tail in mRNAs is taken. Affinity chromatography using oligo(dT) or oligo U as ligand is used for this purpose. Oligo(dT) cellulose is commercially available for the affinity chromatography and fractionation of mRNA is done in the following steps:

1. A column of Oligo(dT) cellulose is prepared using binding buffer containing 10 mM tris, pH 7.5; 0.5 M NaCl, 1 mM EDTA , 10% SDS (added in the buffer just before use) and is equilibrated it using the same buffer.
2. Total RNA isolated is dissolved in the same buffer used for packing the column as described in step 1. Generally, 1 mg/ml solution of RNA is prepared.
3. The RNA solution is applied on the oligo(dT) cellulose column and the column is washed with the binding buffer until absorbance of the effluent is zero. During the process, polyA containing mRNA remains bound with the column matrix and tRNA and rRNA are washed out.
4. Afterwards, elution buffer (10 mM tris, 1 mM EDTA, 10% SDS, pH 7.5) is passed through the column and fractions (1 ml) are collected.
5. The fractions showing absorbance at 260 nm are pooled and mRNA is precipitated out using ethanol as described above. Affinity matrix may be reused repeatedly after regeneration.

Nowadays, ready-made kits are available commercially for isolation of DNA and RNA. These are being preferred since there is more yield and time consumed is lesser with these kits.

References

[1] Sambrook, J., Fritsch, E.F. and Maniatis, T. (1989) *Molecular Cloning* Vol. 1, Cold Spring Harbor Laboratory Press, New York.

[2] Sawhney, S.K. and Singh, R. (2000) *Introductory Practical Biochemistry*, Narosa Publishing House, New Delhi.

[3] Fiske, C.H. and Subbarow, Y.C. (1925) The colorimetric determination of phosphorus. *J. Biol. Chem.* **66**: 375-400.

Chapter II

General Characteristics of the Vectors

2.1. Introduction

Vector is a general term used for bacterial plasmids as well as phage vectors. Vector is defined as replicon which is stably inherited in an extra chromosomal state. Vectors are used for the transformation of cells in gene manipulation. Small plasmids, viruses and phages are vectors since these are replicons. Artificial vectors are constructed by cutting and joining DNA molecules from different sources using various restriction endonucleases and DNA ligase. All the extra chromosomal nucleic acid molecules are not necessarily plasmids or vectors. Vectors imply genetic homogeneity, constant monomeric unit size and the ability to replicate independently of the chromosome. Thus the heterogeneous circular DNA molecules which are found in Bacillus are not necessarily vectors. The definition of vectors include prophages of temperate phages like P1 which are maintained in an extra chromosomal state in contrast to those such as lambda which are maintained by integration into the host chromosome. It also includes coliphages which continuously produce and release phage particles without concomitant cell lysis. Plasmid is a narrow term confined only to the vectors of bacterial origin.

Plasmids may be considered of two different types: conjugative and non-conjugative depending on whether they carry a set of *tra* genes that promote bacterial conjugation. Nowadays, modern plasmids are generally non-conjugative. Modern plasmids lack *mob* gene that is required for mobilization. That's why they are incapable of directing their own conjugal transfer from one bacterium to another. Some old plasmids like pBR322 may be mobilized if a third plasmid (col K) is present in the cell. In fact, col K codes for a mobility protein that nicks pBR322 at a site called as *nic* which is close to the *cis* acting element, *bom*. Mobilization of the plasmid occurs from the nicked site present at position 2254 in pBR322. New plasmids lack *nic/bom* site and that's why are not mobilized.

Plasmids may also be classified as **relaxed plasmids** which are maintained as high copy number per cell and **stringent plasmids** which are maintained as low copy number per cell. The replication of stringent plasmids is coupled to chromosomal replication, hence have low copy number. Copy number of the plasmids is also related to *rop* gene. *Rop* gene codes for rop protein which has negative regulation on replication. Plasmids having mutated *rop* gene may have more copy number. Rop protein also controls compatibility of the plasmids. Rop protein determines whether two different plasmids will coexist in the same bacterial cell. Plasmids utilizing the same replication system can not coexist stably and are said to be incompatible.

2.2. General Characteristics Inserted in the Construction of the Vectors

The following characteristics are inserted during construction of the plasmid vectors:

i. **Selectable markers:** In the construction of bacterial plasmid, generally one or more selectable markers are inserted. Commonly used selectable markers are antibiotic resistant genes and they include:

   - **Tetracycline resistant gene (tet^r):** The gene codes for a protein of 399 amino acids which remains associated with the membrane. The protein prevents the antibiotic tetracycline from entering the bacterial cell.
   - **Ampicillin resistant gene (Amp^r):** The gene codes for an enzyme protein that is secreted into the periplasmic space of the bacterium where it catalyzes hydrolysis of the beta lactam ring of the antibiotic ampicillin thus destroying the antibiotic.
   - **Chloramphenicol resistant gene (Cm^r or Cat):** The gene codes for a tetrameric cytosolic protein having a subunit molecular weight, 23,000. The protein is an enzyme, chloramphenicol acetyl transferase which catalyzes the acetylation of chloramphenicol (formation of hydroxy acetoxy derivatives of chloram) in the presence of acetyl CoA. Hydroxy acetoxy derivatives of chloramphenicol are unable to bind the ribosomes. The expression of chloramphenicol acetyl transferase is catabolite sensitive. There is 5 to 10 fold increased production if the bacteria is grown in a medium other than glucose.
   - **Kanamycin/ neomycin resistant gene (Kan^r):** The gene codes for an enzyme protein, aminoglycoside phosphotransferase. The molecular weight of the enzyme protein is nearly 25,000. The enzyme protein gets located in the periplasmic space. It causes phosphorylation of these antibiotics in the presence of ATP, and phosphorylation of these antibiotics interferes with their active transport into the cell.

ii. **Size:** It is always preferred to keep smaller size of the plasmid. Sequences not necessarily needed for replication of the plasmid are not preferred to be inserted in the plasmid structure. Smaller size plasmids are easy to transform in the bacterial cell compared to larger size plasmids.

iii. **Cloning sites:** If the recognition sequence of any restriction endonuclease occurs once in the entire sequence of the plasmid, it may be considered as the cloning sites.

Now, all the cloning sites are inserted at one place in the vector and that region is called polycloning site region or multiple cloning site region. Earlier, cloning sites were being inserted within the antibiotics resistant genes. Nowadays, it is not preferred since after cloning a DNA fragment in that site, the vector becomes sensitive to that antibiotic.

iv. **Histochemical identification:** Generally construction of the vector is done in such a way that after recombination (cloning of a DNA fragment), it may be distinguished histochemically. For histochemical identification, a common test is of color difference due to β galactosidase. The test was first developed by Messing and coworkers. Wild type *E. coli* cells contain functional β galactosidase which is capable of hydrolyzing a synthetic chromogenic substrate, 5-bromo,4-chloro,3-indolyl-β-D-galactoside, commonly called as X-gal. It is colorless itself. However, after hydrolysis by β galactosidase gives a blue colored product. Functional β galactosidase exists as tetrameric molecule. They prepared a mutant strain of *E. coli* having mutation in the 5' region of the gene coding for β galactosidase (lac Z gene). Mutated protein is not capable of assembling into tetrameric form and does not show enzyme activity. However, if these subunits associate with an oligopeptide corresponding to the N-terminal portion of the enzyme, the enzyme activity gets restored. This association is called 'α-Complementation'. They also manipulated the vector polycloning site by inserting a portion of *E.coli* lac Z gene in such a way that the vector will produce the oligopeptide required for α-complementation, thus making active β galactosidase in the transformed cell. Under the conditions, the colony will be blue colored. However, if any DNA fragment is inserted in any of the cloning site present within the polycloning site, the required oligopeptide will not be synthesized and recombinant transformed colony will be β galactosidase negative and will remain white in color. The same principle is used in many vectors for histochemical identification.

v. **Origin of Replication:** In contructing vector, at least one origin of replication (*Ori* site) is incorporated. Origin of replication is the site from where replication is started. Mostly, *Ori* of *E. coli* is incorporated during construction of the vector. Without, origin of replication, vector will not be able to replicate.

vi. **Promoters:** Promoter site is also incorporated during construction of the vector. Promoter site is important due to the reason that RNA polymerase binds on the promoter site for transcription. If the vector has weak promoter, it will not be able to express cloned DNA fragment. Nowadays, strong promoters are incorporated to make the vector as expression vector. Some scientists have used fused promoters. They fused the strong promoters for incorporation. Some vectors have *tac* promoter which is a fused promoter of *lac* and *trp* promoters. In some shuttle vectors, more than one origin of replication and promoters are incorporated, one for expression in *E.coli* and another for expression in an eukaryotic cell. Nowadays, plasmids/vectors are preferred that allow direct selection of the recombinant clones.

For naming the plasmids/vectors, there is no specific rule. For plasmids, generally p is the first letter indicating plasmid. After that, few scientists prefer to use initials of their own name while others may prefer to use the name of the Institution. For example, in pBR plasmids, B and R are the initials of the workers (Bolivar and Rodriguez). In pUC plasmids, UC are initials for the University of California. Number to the plasmid is given arbitarily by the worker who constructed it. In lambda vectors like λgt vectors, gt indicates generalized transducing. M13 mp vectors are derived from M13 phage.

2.3. Plasmid pBR322

pBR322 is one of the oldest plasmids having 4362 bp. It is a relaxed plasmid having ColE1 replicon. It has two antibiotic resistant genes, Tet^r and Amp^r genes. There are number of cloning sites (single sites) like Ava I, Pst I, BamHI, Pvu II, Cla I, Sal I, EcoRI, Hind III. Ava I site is present at position 1424 (GTCGGG), Pvu II at position 2065 (CAGCTG), Cla I at position 23 (ATCGAT), EcoRI at position 4360 (GAATTC), Pst I at position 3608 (CTGCAG), Hind II / Sal I at position 650 (GTCGAC), BamHI at position 375 (GGATCC), Hind III at position 29 (AAGCTT) are located in the pBR322 sequence. Some single sites are present within the Amp^r gene and some single sites are also present within Tet^r gene. With in Amp^r gene are Pst I, Pvu I and within Tet^r gene are BamHI, Hind II, Sal I restriction endonuclease recognition sequences. The genetic map of pBR322 as line diagram is shown in Figure2.1. Complete nucleotide sequence of the plasmid was determined by Sutcliffe in 1979. Here, if a DNA fragment is cloned in a cloning site within the Amp^r gene, then after cloning, the transformed cell will no more be ampicillin resistant and similarly if the cloning of the DNA fragment is carried in a single site within the Tet^r gene, the transformed recombinant will no more be tetracycline resistant. Here for screening of the recombinants, negative screening approach is done. From the grown colonies on an agar medium containing the antibiotic other than with in whose resistant gene, the DNA fragment cloning in a cloning site is tried, replica plates are prepared. One plate containing the same antibiotic which is used in the grown colonies and the other plate contains the other antibiotic whose resistant gene cloning site is used for cloning the DNA fragment. If the recombinant is present, the colony will not be in that plate containing the same antibiotic but will be in the plate containing the other antibiotic. pBR322 vector contains number of Hinf I restriction endonuclease recognition sequences. Therefore, pBR322 after cutting with Hinf I is generally used as standard marker. Hinf I hydrolyzate of pBR322 gives the following size bands on electrophoresis:

Size in Kb
1.611
0.517
0.506
0.396
0.344
0.298

0.240
0.221

0.154

0.075

Similarly, Hae III cut pBR322 gives the following size bands on electrophoresis and this hydrolyzate is also used as standard marker:

Size in Kb
0.59
0.54
0.50
0.46
0.43

0.27
0.23
0.21
0.19
0.18

0.124
0.123

0.10
0.09
0.08

0.064
0.057
0.050

Although, pBR322 is considered to be a relaxed (high copy number) plasmid, it is not that much high copy number plasmid as pUC plasmids. Many plasmids are transmitted to new hosts by conjugation. However, for conjugation, *mob* gene product is required. Most plasmids in use do not have *mob* gene, therefore are unable to be transformed via conjugation. However, pBR322 can be mobilized by conjugation if a third plasmid, Col K is present in the cell. Col K codes for a mobility protein that nicks pBR322 at *nic* site (at position 2254) close to the *cis* acting element, *bom*. Mobilization of the plasmid occurs from the nicked site. Most other plasmids like pUCs lack *nic*/*bom* site and that's why, these can not be mobilized.

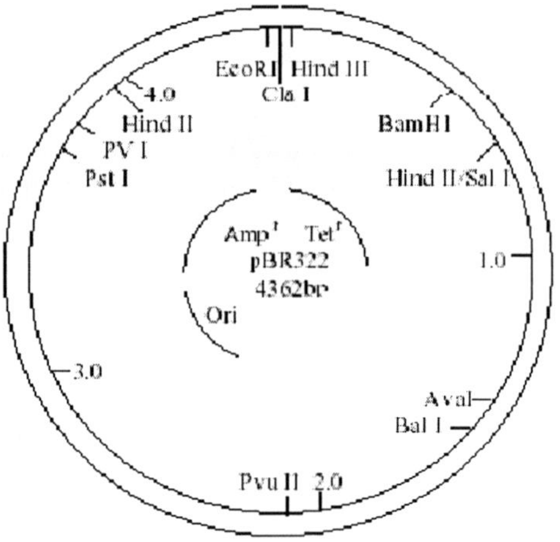

Ori-Origin of replication; Amp^r-ampicillin resistant gene carrying single restriction sites Pst I, Pvu I; Tet^r-Tetracyclin resistant gene carrying single restriction sites; Bam H1, Sal I.
Figure 2.1. Genetic map of pBR322

2.4. pUC Vectors

pUC vectors are derivatives of pBR322 vector. These were developed by J. Messing and coworkers. Here, UC denotes University of California where Messing and Group constructed these vectors. These are relaxed plasmids having ampicillin resistant gene. The genetic map of pUC18 as line diagram is shown in Figure 2.2. *Rop* gene codes for rop protein which reinforces the negative regulation of replication. pUC plasmids have mutated *rop* gene and that is the reason that they have more copy number. At that time, the main characteristic feature of the plasmid was considered to be the presence of multiple cloning site, also called polycloning site and presence of certain sequences having property of α- complementation. Suitable *E. coli* hosts were developed having defective β-galactosidase enzyme gene. Mutations in the 5' region of *lac Z* gene (gene coding for β-galactosidase enzyme) were carried out so that it produces the subunits not capable of assembling into tetrameric form. However, if these subunits associate with a short polypeptide corresponding to amino terminal region of β-galactosidase enzyme, *E.coli* cell is capable to restore the enzyme activity. This association is referred as α-complementation. The enzyme protein synthesized from the mutant gene lacks 11 to 41 amino acids of the normal wild type enzyme protein and is capable of α-complementation. The *E. coli* host has F episome. For the construction of the host, genes responsible for proline synthesis were deleted from the genome and inserted into F episome. Besides, *lac Z* gene was also deleted from the host genome and the mutant form of the gene was inserted into the F episome. The F episome also carries the lac I^q mutation which results in more production of the repressor. This allows the expression of the lac operon to be regulated by the addition of isopropylthio-β-D-galactoside (IPTG), an inducer of the β-galactosidase enzyme. Multiple cloning site having recognition sequences of restriction endonucleases was inserted in such a way that in case of insertion of a DNA fragment in any of the recognition sequence results in the formation of polypeptide incapable of carrying α-complementation. Therefore, histochemical screening of the recombinant pUC vectors can be done. If 5-bromo-4-chloro-3-indolyl-β-D-galactoside, commonly called as X-gal, a chromogenic substrate for β-galactosidase enzyme is used, recombinant vector containing *E. coli* colony will be white (colorless) since it will have inactive β-galactosidase enzyme and blank (without cloned fragment) vector containing colony will be blue colored. Sigma-Aldrich recently developed a new chromogenic substrate, 3,4-cyclohexanoesculetin-β-D-galactopyranoside (S-Gal, a registered trade name of Sigma-Aldrich) for β-galactosidase in place of X-gal. S-gal is autoclavable/microwavable and is more soluble in water compared to X-gal. After hydrolysis of S-gal by β-galactosidase, the hydrolyzed aglycone (non-sugar portion) reacts with the Fe^{+3} of added ferric ammonium citrate to produce an intense black stain. This stain provides enhanced contrast.

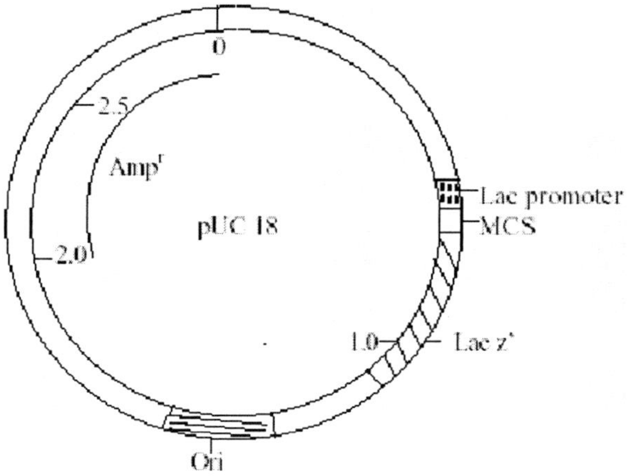

Ori- Origin of replication; Amp^r- ampicillin resistant gene; MCS-multiple cloning sites; Lac Z'-portion of βgalactosidase gene

Figure 2.2. Genetic map of pUC18 vector

For construction of pUC plasmids, a portion of pBR322 from 2067 to 4362 was taken which has ampicillin resistant gene and was called pUR1. Afterwards, the Pst I site present in position 3612, CTGCAG was blocked by changing to CTACAG and the resultant plasmid was called pUC3. Afterwards, Hinc II site present in position 3908, GTCAAC was blocked by changing to GTCAAT and the resultant plasmid was called pUC5. Afterwards, Acc I restriction endonuclease site present in position 2245,GTATAC, was deleted using BAL31 nuclease resulting in pUC6.

Hae II fragment from 804 to 1237 was cut from M13mp7 and was inserted into the Hae II site present at position 2352 of pUC3 to construct pUC4. and of pUC6 to yield pUC7.

With in multiple cloning site, recognition sequences of EcoRI, Bam HI, Sal I, AccI, Hinc II, Pst I are present in case of pUC7. In pUC8, recognition sequences of EcoRI, Sma I, Xma I, Bam HI, Sal I, Acc I, Hinc II, Pst I, Hind III are present with in the multiple cloning site. In pUC9, recognition sequences of Hind III, Pst I, Sal I, Acc I, Hinc II, Bam HI, Sma I, Xma I, EcoRI, Hae II are present with in the multiple cloning site. The sequences in pUC7, pUC8 and pUC9 are such that these three are in three different reading frame. If a DNA fragment is inserted in a restriction site in all the three plasmids, it will be aligned in a proper reading frame in one plasmid. pUC12 and pUC13 are identical except these have multiple cloning site in reverse orientation of each other. Restriction sites of EcoRI, Sst I, Sma I, Xma I, Bam HI, Xba I, Sal I, Acc I, Hinc II, Pst I, Hind III are present in the multiple cloning site of these two plasmids, pUC12 and pUC13. Similarly, pUC18 and pUC19 are identical except they have multiple cloning site in reverse orientation of each other. Restriction sites of EcoRI, Sst I, Kpn I, Sma I, Xma I, Bam HI, Xba I, Sal I, Acc I, Hinc II, Pst I, Sph I, Hind III are present in the multiple cloning site of pUC18 and pUC19. The sequence of the multiple cloning sites of pUC vectors is given in the Figure. Average size of pUC vectors is nearly 2.7 Kb. For example, size of pUC19 is 2686 bp. These are able to clone the fragments of nearly 2-3 Kb.

In contrast to pBR322, in pUC vectors, cloning sites are not present with in the antibiotic resistant gene. Therefore, host colonies either having blank or recombinant pUC vector remain ampicillin resistant. Under the conditions, there will not be growth of the untransformed colonies.

2.5. M13 Vectors

M13 vectors have been derived from the bacteriophage M13. These were developed by Messing and coworkers. M13 vectors have generally been used for dideoxy DNA sequencing method and for oligonucleotide site directed mutagenesis where is use of a single stranded vector. In fact, M13 vectors were constructed by Messing and group considering the need of single stranded vector. M13 is a male specific filamentous bacteriophage of *E. coli*. The phage particle contains a single stranded circular DNA and an outer protein coat. It attaches to the F pilus for infection. Therefore, M13 phage/ vectors do not infect *E. coli* cell not containing F pilus. After infection of the *E. coli* cell, M13 (+) strand DNA serves as a template for the synthesis of the complementary (-) strand DNA with the formation of a double stranded replicative form of the phage DNA. The double stranded DNA gets amplified till there are about 200 copies per cell. Afterwards, there is not amplification of the (-) strand. Only (+) strand DNA continues to be synthesized. As the (+) strand DNA is synthesized, it displaces the previous (+) strand DNA. The dissociated (+) strand DNA gets packaged in the phage coat proteins to form viral particle and (-) strand DNA continues to act as a template for more synthesis of the (+) strand DNA. After the formation of the phage particle, it comes out of the bacterial cell without rupturing it. M13 phage is neither a lysogenic nor lytic type phage. It does not form a true plaque on a bacterial lawn due to its non-lytic behaviour. This property of M13 was considered by Messing and coworkers for constructing the vectors. M13 phage is a filamentous phage, therefore, it does not have tight constraints on the size of the DNA that can be packaged efficiently. M13 vectors have the advantage that larger DNA fragments (upto 7-8 Kb size) may be cloned in these vectors. Besides, double stranded form of the vector has about 200 copies per cell, therefore, may be isolated and purified easily for use as cloning vector. The M13 particles may be precipitated out easily from liquid culture by adding polyethylene glycol/ sodium chloride after the removal of *E. coli* by centrifugation.

For construction of the vector, M13 DNA was modified and use of α-complementation was taken. Functional β-galactosidase exists as a tetrameric protein in *E. coli*. The enzyme is able to convert the synthetic colorless substrate, 5-bromo, 4-chloro,3-indolyl-β-D-galactoside (commonly called as X-gal or Bluogal) to a blue colored chromophor, allowing its visual detection. Mutation in the 5'-region of the gene coding for β-galactosidase (lac Z) produce subunits that are not capable of assembly and do not show enzyme activity. However, if the subunits associate with a short polypeptide corresponding to the N-terminal portion of the enzyme, β-galactosidase enzyme activity gets restored. This association is called as α-complementation.

The first modification in the M13 DNA was done by inserting a portion of *E. coli lac* operon into the intergenic region of M13. This DNA sequence was capable of coding for a peptide capable of α-complementation. This vector was called M13mp1. Messing and coworkers also manipulated *E. coli* so that it could work as host for M13 vectors. The *lac* operon was deleted from the *E. coli* genome and the mutated form of the gene (*lac* Z ΔM15) inserted into the F episome. The mutated *lac* Z gene codes for β-galactosidase protein lacking 11-41 amino acids of the wild type protein but is capable of doing α-complementation. The F episome also carries the lac I^q mutation which results in the overproduction of the *lac* cytoplasmic repressor. This allows the regulation of the expression of the *lac* operon by the addition of isopropylthio-β-D-galactoside (commonly called as IPTG), *lac* operon inducer. It is essential for M13 infection that its host strain carry the F episome. The genes responsible for proline biosynthesis were deleted from the *E. coli* genome and inserted into the F episome. Under the conditions, *E. coli* cells can be maintained on minimal media with the retention of the F episome. Under the conditions, the uninfected host produce inactive β-galactosidase. Upon infection with M13mp1, these cells produce active β-galactosidase due to α-complementation. Subsequently, blue plaques are obtained in the presence of X-gal.

The next step in the construction of M13 vectors was the insertion of the recognized sites (recognition sequences) for different restriction endonucleases into the region of lac Z gene in M13mp1. The same recognition sequences were inserted and the vector was same numbered as in pUC vectors described above. The nucleotide sequence of multiple cloning sites present in various pUC/M13mp vectors is shown in Figure 2.3. It is pertinent to mention that in both the cases, since these sites were considered as cloning sites, any recognition sequences of any of the restriction endonuclease present in the multiple cloning site region if present in the vector was blocked by mutation. These sites were inserted such that the added nucleotides did not interfere with the capability for maintaining proper reading frame. Insertion of DNA fragments into any of the recognition sites results in a peptide incapable of carrying α-complementation. As a result, plaque will be colorless. Near the multiple cloning site, the sequences were inserted such that there should be annealing of the universal primer (for use in dideoxy DNA sequencing). Therefore, recombinant and blank plaque may be distinguished on the basis of blue white screening of the plaques as described in case of pUC vectors. It is pertinent to mention that M13 vectors do not have any antibiotic resistant gene. Average size of M13mp vectors is nearly 7.25 Kb. For example, size of M13mp18 is 7249 bp. On an average, fragments of up to 6-7 Kb may be cloned in M13 vectors without difficulty. Although nowadays with the help of polymerase chain reaction vector DNA can be amplified, the classical method of amplification of the M13 vector DNA is as follows:

M13mp plaque is picked up with the help of a Pasteur pipet and is transferred in to 2 ml of a suitable media (for example 2x YT medium) containing *E. coli* culture. For the purpose, an overnight culture after dilution 100x with 2x YT medium may be used, and allowed to grow for 6-8 hours. During this period, M13 plaque will infect the *E. coli* cells, multiply inside the cell and as a result as described above will form approximately 200 copies of the replicative form and afterwards, only (+) strand will be amplified and after covering with the protein coat, viral particles without rupturing the *E. coli* cells will come out in the medium. Overnight culture of *E. coli* after infection with M13 is avoided otherwise there may be rupturing of the *E. coli* cells and medium will also contain genomic DNA of *E. coli*. After 6-8 hours of incubation at 37°C, the suspension is centrifuged after transferring in an eppendorf tube. The pellet is *E. coli* cells having replicative form of the M13 vector which may be used for isolation of replicative form of the vector for cloning purpose using the protocol of isolation of the plasmid DNA. The supernatant will have M13 viral particles. Although it is not preferred to discontinue the protocol, if necessary, it may be discontinued at this stage and if so supernatant should be stored at 0-4°C in a refrigerator. Next morning when it has to be started, the supernatant should be centrifuged before proceeding further since it is possible that few cells of *E. coli* may be present in the supernatant and these might have multiplied during storage although at a slow rate since the storage was in the cold condition. To the supernatant (1.2 ml), 0.25 volume of PEG/NaCl reagent (20 % polyethylene glycol (PEG) and 2.5 M NaCl) is added, mixed and is stored at the room temperature for 15 minutes. Afterwards, centrifuge for 10 minutes to pellet the virus. PEG/NaCl precipitates the viral particles. Supernatant is removed carefully without disturbing the pellet and is discarded. Pellet is suspended in 100 μl of TES buffer (20 mM tris, 10mM NaCl, 0.1 mM EDTA, pH 7.5). This suspension is of viral particles. This suspension may be used for direct infection of the *E. coli* cells. The suspension is extracted with equal volume of Phenol-chloroform-isoamylalcohol (25: 24: 01) to extract proteins. Upon extraction, phenol layer being heavy would be lower phase and aqueous phase having DNA will be the upper phase and at the interface will be denatured proteins. The aqueous layer is separated carefully with the help of a pipet without disturbing the interface and is transferred in an another eppendorf tube and the process is repeated at least once although preferred twice. The aqueous layer is extracted either with diethyl ether or chloroform to remove traces of phenol if any in the preparation. Afterwards, the preparation is incubated at 37°C for few minutes for evaporation of the traces of diethyl ether or chloroform if any. To the preparation, 2.5 M sodium acetate, pH 4.8-5.2) is added to get final concentration 0.25 M and 2 to 3 volumes of chilled ethanol are added, mixed and incubated in dry ice- ethanol for 15-20 minutes to precipitate DNA. Precipitation of DNA may also be carried out by incubation at –20°C overnight. Afterwards, the contents are centrifuged in cold condition to collect the pellet of DNA and the pellet is washed with chilled 80 % ethanol. The pellet which is invisible is dried under vaccum generally using a centrifugal lyophilizer and is dissolved in TE buffer or water.

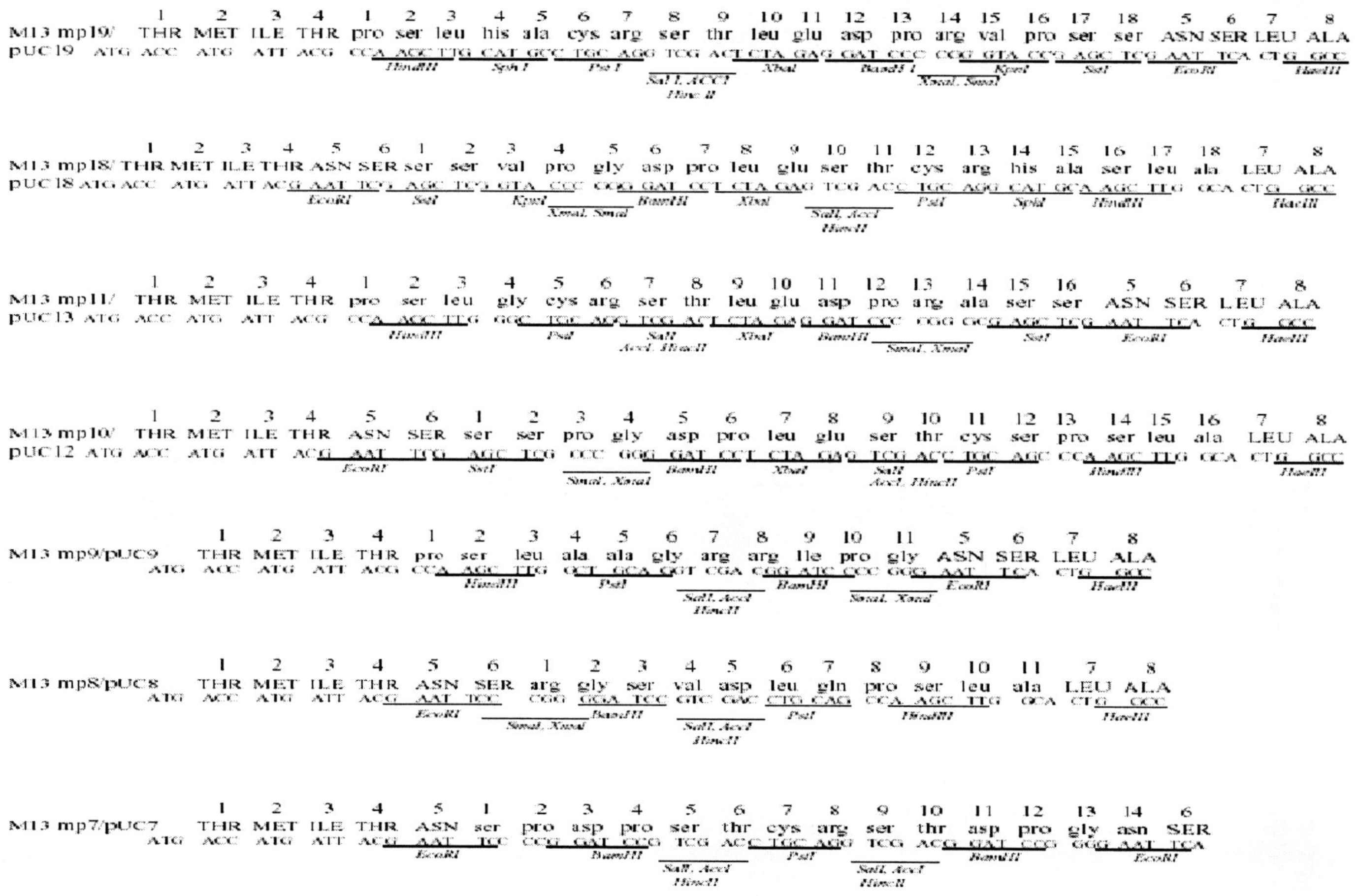

Figure 2.3. Sequences of the multiple cloning sites in M13mp/pUC Vectors

2.6. EMBL Vectors

Although pUC and M13mp vectors are versatile because they allow one to insert foreign DNA fragments at a number of restriction endonuclease sites with in a short polylinker DNA contained in the coding sequence of the α-peptide of the enzyme β-galactosidase. Besides, colonies/plaques may also be screened by simple color tests without preparing any replica plates. In order to make these vectors more versatile, Dente et. al. constructed a new series of vectors by inserting 1300 bp EcoRI-EcoRI segment of the fl genome from the plasmid pD4 into the unique Nar I site present in all pUC vectors. The resulting vectors named as pEMBL vectors contained all the pUC specific characters including the same origin of replication and Amp^r (ampicillin resistant gene). The pEMBL vectors are also able to synthesize α-peptide of the enzyme β-galactosidase and so produce blue colonies when plated on the appropriate media in the presence of chromogenic substrate, X-gal. Besides, as in pUCs, the gene for the α-peptide is inactivated in recombinant plasmids resulting in the formation of white colonies on plates containing X-gal. Since the pEMBL plasmids contain the intact origin of replication of fl that can be activated when cells are superinfected with fl phage. As a result, there is production of single stranded plasmid DNA that may be packaged and secreted into the medium as a virus like particle. The orientation of the 1300 bp segment of fl DNA determines which of the two strands is encapsidated in the viral particle and accordingly the vectors have been named as pEMBL (+) and pEMBL (-). In the pEMBL (+) vectors, the antisense strand of the β-galactosidase gene is present in the virion produced by superinfected cells. In the pEMBL (-) vectors, the coding strand of the β-galactosidase gene is present in the virion.

2.7. Lambda (λ) Vectors

There are two most commonly used lambda vectors called as λgt 10 and λgt 11. In the nomenclature, gt signifies that such a derivative can act as a generalized transducing phage through the incorporation of bacterial genes by manipulation *in vitro*. These vectors are generally used for constructing cDNA libraries of eukaryotes. Due to the presence of introns in the genome of eukaryotes, genomic library preparation is not preferred. For cDNA library preparation, first mRNA is isolated and with the help of reverse transcriptase enzyme, complementary DNA (cDNA) is prepared and used in the construction of cDNA library. Phage library cloned in λgt 10 may be screened using nucleic acid probe. λgt 10 is considered to be non-expression vector since it does not have a strong promoter upstream of the cloning site. It does not mean that expression of the cloned cDNA can not occur in λgt 10 under any circumstance. If the cloned fragment contains promoter in itself, then the expression of the cloned fragment may occur. λgt 11 is an expression vector capable of expressing the cloned cDNA. Therefore, phage library in λgt 11 may be screened using antibody probe to isolate DNA sequences coding the antigens against which the antibodies are raised.

λgt 10 contains a single EcoRI recognition site within the cI gene. The genetic map of λgt 10 as line diagram is shown in Figure 2.4. Fragments upto about 7.6 Kb size may be cloned in it. EcoRI site present within the cI gene is used as a cloning site, therefore, after cloning of the fragment in EcoRI site, it generates cI^- phage which forms plaque with clear center whereas, phage containing cI^+ (phage without cloned fragment) produces turbid plaque. Therefore, recombinant phage plaque may be screened on the basis of clear or turbid plaque. λgt 10 does not require an insert fragment to yield a packageable DNA molecule, therefore, initially, cDNA library cloned in λgt 10 may have mostly parent phage lacking inserts. Under the conditions, screening a library consisting of mostly parent phages without insert may be troublesome. This problem may be countered by selecting against the non-recombinant parent phage during amplification of the cDNA library. The selection differentiates between cI^+imm^{434} and cI^-imm^{434} phage. If an *E. coli* strain carrying high frequency lysogeny mutation hflA150, is infected by λgt 10 or any other cI^+imm^{434} phage, the cI^+ phage is repressed so quickly that plaque formation is suppressed. However, there is no suppression of cI^- phage. Taking the advantage of this property, cDNA libraries in λgt 10 are plated on hflA host.

In case of λgt 11 (*lac*5, *cI*887, *nin*5, *S*100), a single EcoRI recognition site is present within the *lac* Z gene, 53 bp upstream from the termination codon. The genetic map of λgt 11 as line diagram is shown in Figure 2.5. Fragments upto about 7.2 Kb size may be cloned in it. EcoRI site present within the *lac* Z gene is used as a cloning site. The vector produces a temperature sensitive cytoplasmic repressor (cI887) which becomes inactive above 42°C and contains an amber mutation (*S*100) which renders it lysis defective in hosts lacking the amber suppressor. If a fragment is cloned in EcoRI site, it may be expressed as fusion protein with β- galactosidase. Therefore, the library cloned in λgt 11 may be screened using antibody probe. In a typical experiment, 10^5 to 10^6 individual recombinants as phage plaques may be screened on a lawn of lon protease deficient *E. coli* cells. Proteins released by the lysis of the cells within the plaque are immobilized on a nitrocellulose membrane placed over the lawn. Afterwards, protein bound to the nitrocellulose membrane is probed using antibody specific to the protein. Antibody binding can be detected using secondary antibody tagged either with the radioactive material like I^{125} or with alkaline phosphatase which may be detected using chromogenic substrate.

While constructing λgt 11, attention is paid to the problem of foreign protein expression which may be toxic for the host cell. Since foreign DNA sequence is fused to *lac* Z gene, therefore, it gets expressed efficiently in *E. coli*. The second problem is if the foreign protein is toxic to the host cell and might kill it before sufficient expression takes place. The problem is minimized by controlling the expression of *lac* Z gene by using the host cell producing large amount of lac operon repressor which prevents the expression of the fusion protein during the initial phase of plaque formation. When the number of infected cells surrounding the plaque is sufficient, expression of the *lac* Z is induced by adding IPTG which inactivates *lac* cytoplasmic repressor. Under the conditions, detectable amount of the foreign antigen may be produced even if the foreign protein is toxic to the cells. The third problem is stability of the foreign protein within the host cell. The construction of the vector is such that the foreign gene gets cloned within lac Z gene in such a way that it is present near the carboxyl terminus portion of β-galactosidase which increases its stability. This problem is also overcome by using lon protease deficient cells.

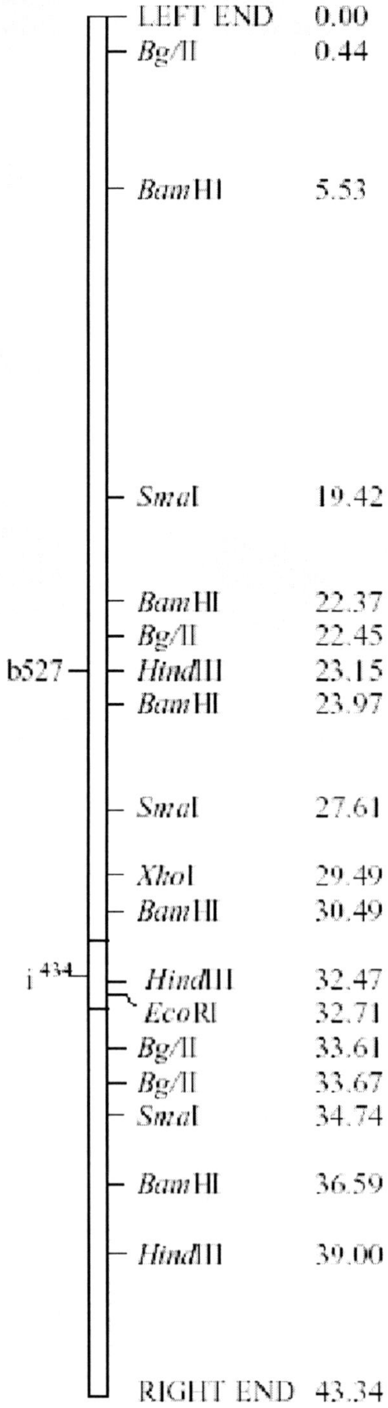

Figure 2.4. Genetic map of λgt 10

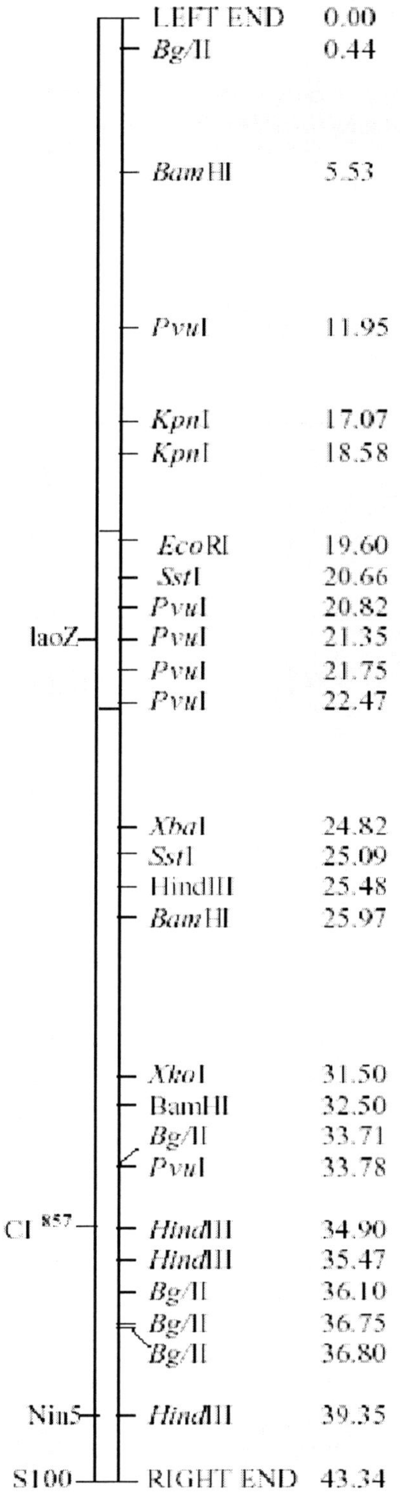

Figure 2.5. Genetic map of λgt 11

Lambda phage may be isolated and purified using the following protocol:

1. Cool the lysed culture to room temperature and add DNase I and RNase each to a final concentration of 1 μg/ml. Incubate for 30 min. at the room temperature. The step is carried out to chew the bacterial DNA and RNA which get released as a result of the lysis of the bacterial cell. Otherwise, at a later stage, bacterial DNA/RNA will contaminate the viral DNA.
2. Add solid NaCl to a final concentration of 1M (58.5 g per litre). Dissolve gently by swirling the flask. Incubate for one hour on ice. The addition of NaCl aids dispersion of the bacteriophage particles from bacterial debris and is needed for efficient precipitation of bacteriophage particles when in later step, polyethylene glycol (PEG) is added.
3. Remove debris by centrifugation at high speed (nearly 10,000x*g*) for 10 min. in the cold condition (0-4°C). The pellet will be bacterial debris which may be discarded. Precautions should be taken before discarding since it may also contain adhered viral particle. It is preferable to autoclave the stuff before discarding.
4. Add solid PEG (MW 8000) to a final concentration of 10% (W/V). Dissolve gently. Chill in ice water and incubate on ice for at least one hour. The process allows the bacteriophage particles to form a precipitate.
5. Centrifuge the suspension at nearly 10,000x*g* for 10 min. in the cold condition to collect the bacteriophage particles in the form of a pellet. Supernatant may be discarded after autoclaving.
6. Suspend the phage pellet in 8 ml SM medium for 500 ml culture. SM medium is NaCl, 5.8 g; $MgSO_4.7H_2O$, 2 g; 1M tris, pH 7.5, 50 ml; 2% gelatin solution, 5 ml; and water to make one litre.
7. Extract PEG and cell debris, if any, from the phage suspension by adding equal volume of chloroform and vortexing is done for mixing. The two layers may be fully separated by centrifugation at 3,000x*g* for 15 min. in the cold condition. Aqueous layer should be transferred into another tube which contains bacteriophage particles.
8. Add cesium chloride to a final concentration of 0.5g/ml. Density gradient centrifugation is carried out at 22,000x*g* for 2 hours in the cold condition. A bluish band of bacteriophage particles should be visible at the interface between the 1.45 and 1.5g/ml layers. If the yield is low, the band may be detected by showing light from above and putting black background.
9. Collect the bacteriophage particles by puncturing the side of the tube. First a piece of scotch tape is put on the outer side of the tube level with the bacteriophage band. Puncture the tube with the help of a 22 gauge needle through the tape and collect the band of the bacteriophage particles. Care should be taken that bacteriophage band should not get contaminated with any other material.

Although bacteriophage particles band obtained in step 9 above may be used to isolate lambda DNA. It is preferable to further purify the bacteriophage particles. For the purpose, add cesium chloride in the bacteriophage suspension to a final concentration of 1.5g/ml.

Centrifuge using Ti type 50 rotor at 38,000 rpm for 24 hours in the cold condition in an ultracentrifuge. Collect the band of the bacteriophage in the same way as described above.

Since lambda particles are much sensitive to EDTA, therefore, it is essential that Mg^{+2} (10-30 mM) be present all the time during purification to prevent disintegration of the particles. If the yield of the phage is low, the number of infectious bacteriophage particles should be determined in samples taken at various stages during the purification in order to determine the losses.

From the isolated lambda phage, DNA may be extracted using the following steps:

1. Remove cesium chloride from the phage suspension by dialysis at the room temperature for 2-4 hours against a 1000 fold volume of 50 mM tris-HCl, pH 8.0 containing 10 mM $MgCl_2$ (TM buffer) with at least one change after one hour or so.
2. Transfer the contents present in the dialysis tube into a tube of suitable size so that it should fill about 1/3rd only.
3. Add 0.5 M, pH 8.0 EDTA to give a final concentration of 20 mM. EDTA is added to rupture the phage particles. As written above, Mg^{+2} are required for keeping the phage particles intact. EDTA chelates Mg^{+2}.
4. Add pronase enzyme to a final concentration of 0.5 mg/ml. Proteinase K may also be used in place of pronase, however, then concentration used is 50 μg/ml. The mixture is incubated at 50°C for few minutes. This is done to chew the proteins.
5. Add sodium dodecyl sulfate (SDS) to a final concentration of 0.5%. Mix well.
6. Incubate for one hour at 37°C, if pronase is used or at 65°C if proteinase K is used.
7. Add equal volume of Phenol/chloroform/isoamyl alcohol (25:24:01) to extract the proteins which get collected at the interface. Details of the extraction are described above.
8. After removing the traces of phenol, perform dialysis overnight in the cold condition against TM buffer.

2.7.1. Lambda ORF8 Vector

It has been derived from lambda phage. Size of the vector is 42.84 Kb. It can be used for construction of cDNA library. Ampicillin resistant gene (Amp^r) and *lac* operon DNA have been incorporated during the construction of this vector. The DNA fragments of nearly up to 9 Kb may be cloned in this vector. There are unique Hind III, Bam HI, and EcoRI sites located with in the *lac* Z coding region. The genetic map of lambda ORF8 as line diagram is shown in Figure 2.6. Therefore, if a cloned DNA sequence is in frame with the *lac* Z sequences, there may be expression of the fusion protein. However, if the cloned DNA sequence is not in frame with the *lac* Z sequences or if termination codons are present, recombinants are expected to be lac Z^-. The right arm contains an amber codon in *int* at codon 300, has the *nin* 5 deletion and is *imm* 21 (*imm* 21 means that substitution from phage 21 is incorporated). It also has substitution of *lac* Z^+, *lac* Y^+, *lac* I^{q1}. Blank as well as recombinant both are red^+, gam^+. (*gam* gene codes for an inhibitor of the *E. coli rec* BC exonuclease, thereby allowing an efficient transition from θ type DNA replication to rolling circle replication is used). Left arm will be nearly 24.7 Kb and right arm 18.1 Kb. It propagates on *rec* A$^-$ hosts.

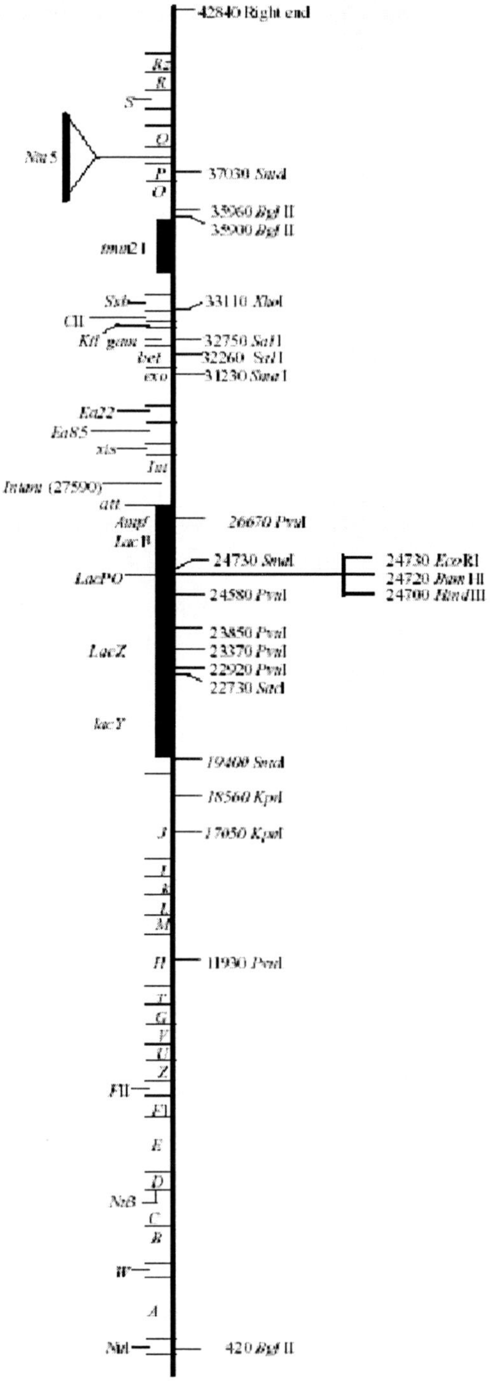

Figure 2.6. Genetic map of λORF 8

2.7.2. Charon Vectors

Charon bacteriophages are replacement vectors designed for cloning large fragments of DNA. Various charon vectors have been constructed. Charon vectors constructed late like Charon 32-35, Charon 40 accept larger DNA fragments and have more useful restriction sites. Stuffer of charon 40 consisted of smaller DNA fragments joined head to tail and can be digested using Nae I restriction endonuclease. General size of the left arm is 48.5 Kb and of the right arm 45.36 Kb.

(a) Charon 4A: It is used as a substitution vector for construction of eukaryotic genomic libraries. There are two EcoRI sites located within the lac Z gene which may be used as cloning site. The genetic map of Charon 4A as line diagram is shown in Figure 2.7. It will be a substitution since a fragment after cut in both the EcoRI sites will be replaced by the cloning fragment. Using EcoRI as cloning site, generally DNA fragments upto 20 Kb (7-20 Kb) may be cloned in it and the recombinant will show Lac^-, Bio^-. Due to removal of the fragment by EcoRI cut, size of the left arm will be 19.5 and of the right arm 11 Kb. The other restriction site which may be used as a cloning site is Xba I where insertion will take place since there will not be removal of any fragment. Using Xba I, fragments of upto 6 Kb may be cloned in it and recombinant will be Lac^+, Bio^+. Here, the left arm will be 24.8 and right arm 20.5 Kb.

The amber mutations in genes A and B (Aam 32, Bam 1) allow this vector to be used for screening recombinant libraries by *in vivo* recombination. The amber mutations require Sup E &/or Sup F hosts for propagation. The host should also be rec A^+ since the vector is red^-, gam^-. In the construction of the vector, a deletion in the xX-cI region (KH 54) is carried out that prevents lysogenization. It also has *nin* 5 (deletion of tR_2 rendering delayed early transcription independent of the N gene product). Substitution of lac 5, bio 256 and QSR 80 has also been done.

(b) Charon 32: It is used as a substitution as well as insertion vector for construction of eukaryotic genomic libraries. There are two EcoRI sites which may be used for cloning 8 to 19 Kb DNA fragments. The genetic map of Charon 32 as line diagram is shown in Figure 2.8. Under the conditions, it acts as substitution vector. Due to removal of the fragment, left arm size is 19.5 Kb and right arm size 11.9 Kb. Similarly, two Hind III restriction sites present in the vector may be used for cloning 5 to 18 Kb size DNA fragments and the vector acts as substitution vector. The size of the left arm will be nearly 20 Kb and of the right arm nearly 13.3 Kb. There is one Sac I site which may be used as a cloning site to insert fragments of upto 10 Kb size. Under the conditions, it acts as insertion vector. The size of the left arm will be 27 Kb and of the right arm 16.8 Kb. It propagates on *rec* A^- host. In the vector, there is substitution of a 12.5 Kb mouse stuffer. Deletions are KH 54 and nin 5. The vector as well as recombinants are red^-, gam^+.

(c) Charon 40: It is used as a substitution vector for the construction of genomic libraries. It has been constructed in such a way that it may clone the DNA fragments of 9.2 to 24.2 Kb. There are many restriction sites present in inverted polycloning sites at opposite ends of the polystuffer. The genetic map of Charon 40 as line diagram is shown in Figure 2.9. The combinations of the restriction sites which may be used for substitution of the cloning fragment are EcoRI-Apa I, Xba I-Sfi I, Sac I-Avr II, Sal I-Spe I, Hind III-Xho I, Bam HI-Nae

I, Kpn I- Not I, Sma I-Xma III. It has WL 113, KH 54, nin 5 deletions. It has lac operator and adenovirus DNA fragment, nearly 80 times as substitution. Due to substitution, the size of the left arm will be 19.2 Kb and of the right arm 9.6 Kb. It propagates on *rec* A$^-$ host. The vector as well as recombinants are red$^-$ gam$^+$.

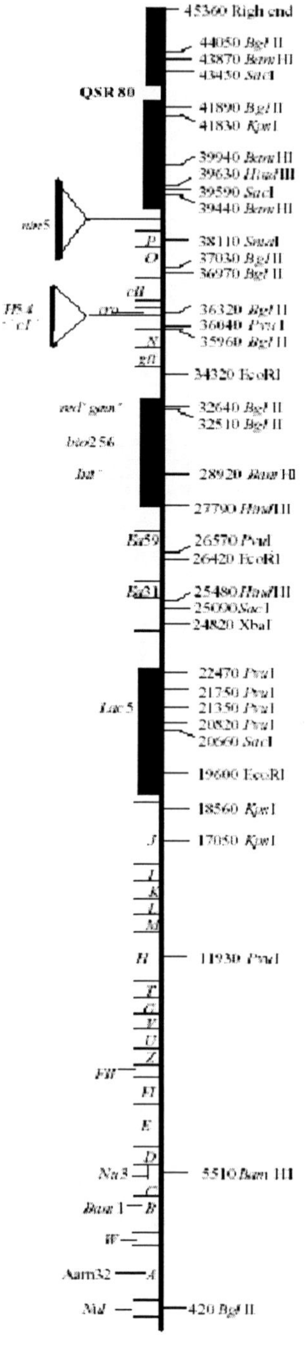

Figure 2.7. Genetic map of Charon 4A

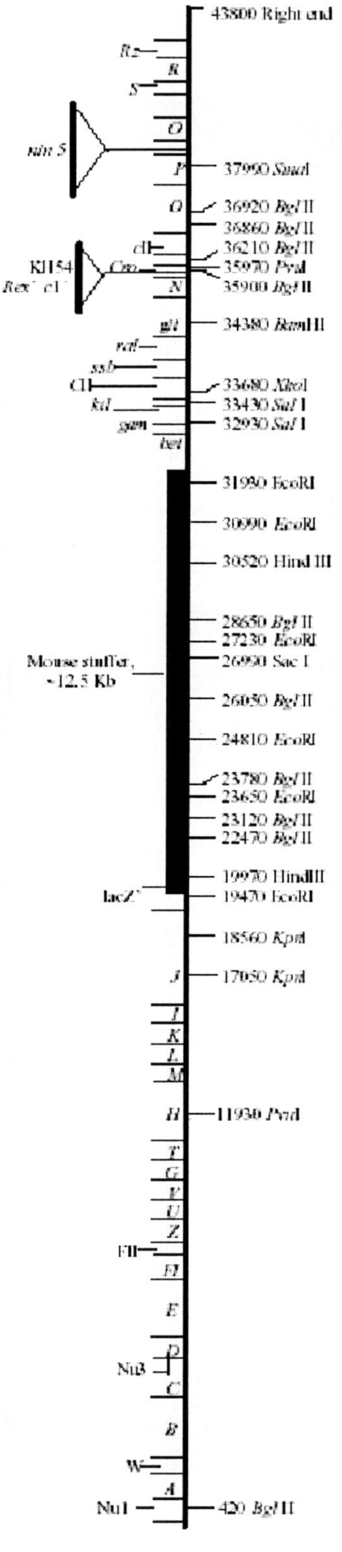

Figure 2.8. Genetic map of Charon 32

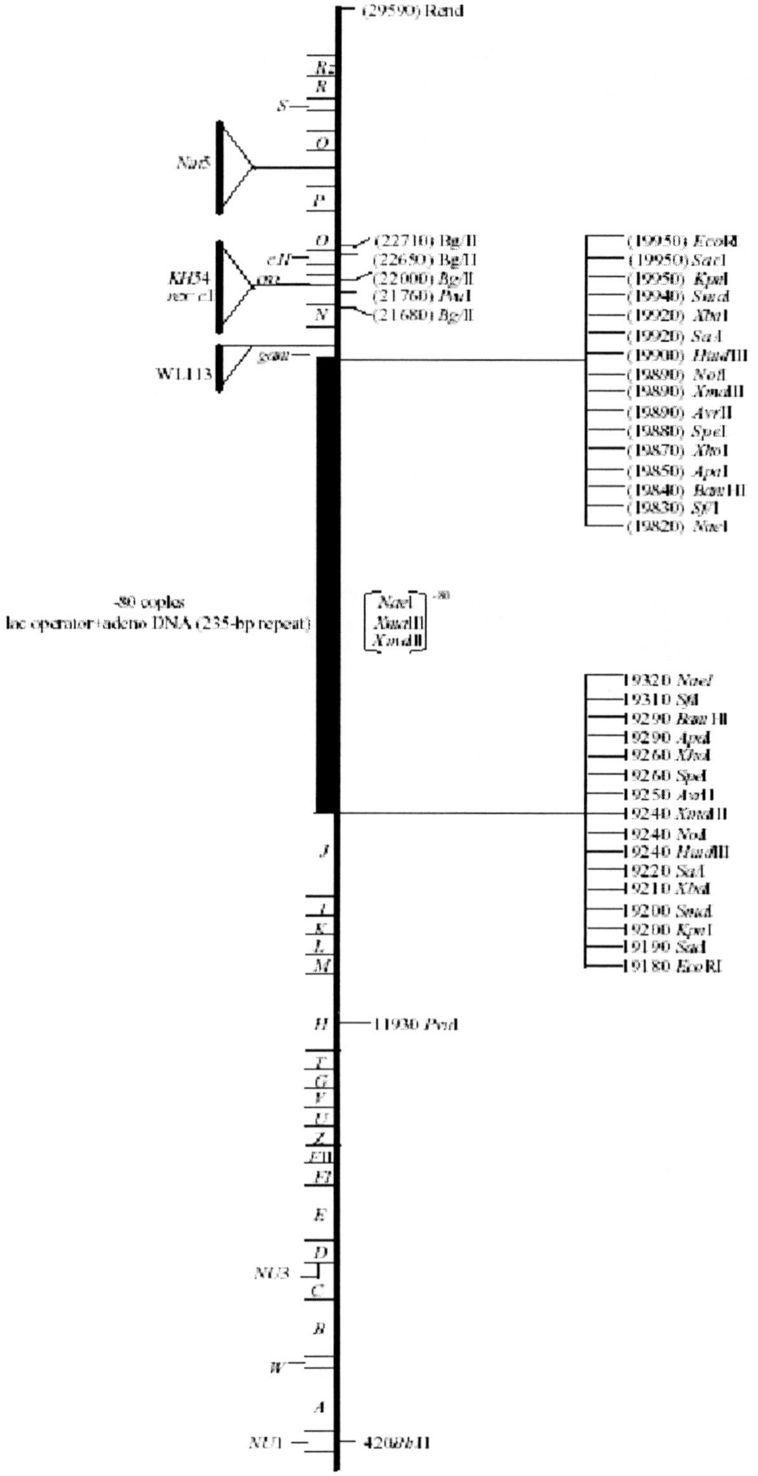

Figure 2.9. Genetic map of Charon 40

2.8. Cosmid Vectors

These are the modified plasmids that carry at least one cos site (DNA sequences required for packaging DNA into bacteriophage λ particles. They carry an origin of replication, usually of col E1 origin) and a drug resistant marker, usually Amp^r. Cosmid vectors may be introduced into *E. coli* using standard transformation protocol. Cosmid vectors construction is in such a way that these are able to clone the DNA fragments of even more than 40 Kb size. Therefore, mostly cosmid vectors are used for constructing eukaryotic genomic libraries.

(a) pJB 8: It is a cosmid vector of about 5.4 Kb size. It contains an ampicillin resistant gene, Amp^r, a col E1 origin of replication, a single cos site derived from charon 4A and four restriction sites, namely Bam HI, Eco RI, Cla I, Hind III that can be used for cloning. The genetic map of pJB8 as line diagram is shown in Figure 2.10. It is capable of cloning DNA fragments of 33 to 46.5 Kb. However, it is not a vector of choice due to the problem of concatenation due to the presence of only one cos site. Generally, cosmid vectors having two cos sites are recommended for cloning.

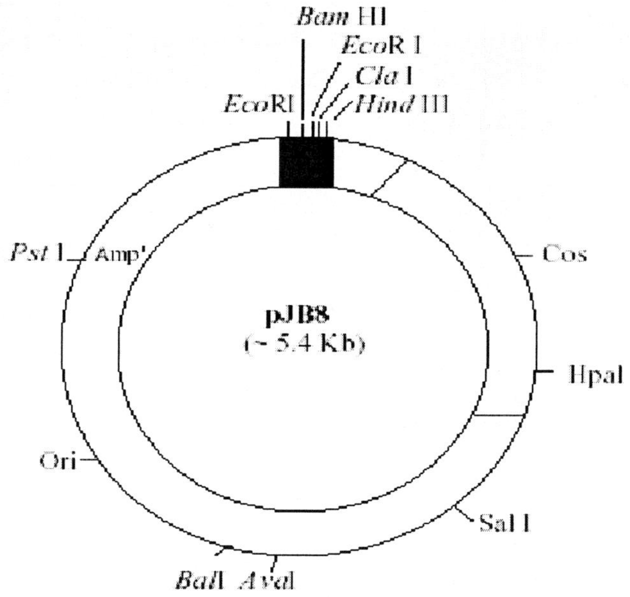

Figure 2.10. Genetic map of pJB8

(b) C2RB: It is a cosmid vector of about 6.8 Kb size. It contains Amp^r and Kan^r genes and col E1 origin of replication derived from pKC 7 plasmid. The genetic map of C2RB as line diagram is shown in Figure 2.11. It has two cos sites derived from MUA-3 vector. There are four restriction sites, namely Bam HI, Eco RI, Cla I, Hind III that can be used for cloning. It is capable of cloning DNA fragments of 33 to 46.5 Kb. There is loss of 1.7 Kb fragment containing a cos site during packaging. This 1.7 Kb fragment also contains Kan^r gene, therefore, kanamycin resistance character is lost during packaging. For cloning DNA fragment, vector is sequentially digested with Sma I and a second restriction site (out of the four mentioned above) for example Bam HI. This sequential digestion creates two fragments of the vector, each carrying a cos site and each having a cohesive end and a blunt end. The digested vector DNA is then ligated to phosphatase treated segments of foreign DNA of the appropriate length carrying termini compatible with the cohesive termini created by Bam HI. It has been found that presence of high concentrations of ATP (say like 5 mM or more) in the ligation mixture inhibits ligation of the blunt ended DNA without affecting the ligation of the cohesive termini.

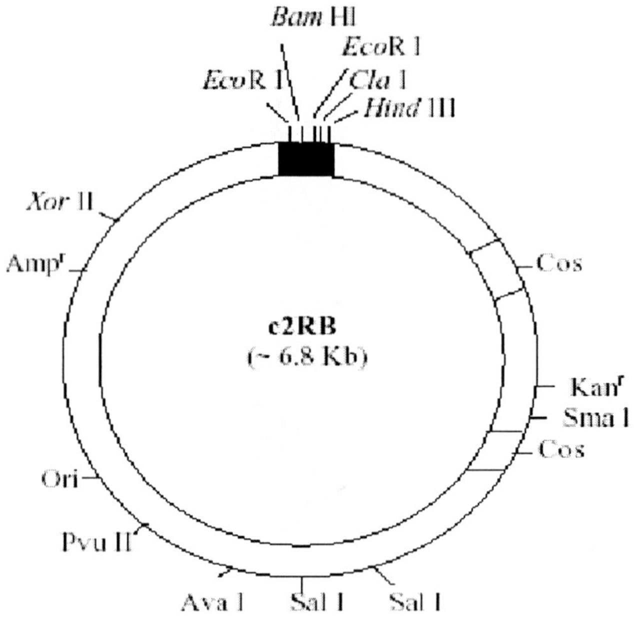

Figure 2.11. Genetic map of c2RB

Cosmic Vectors for Transformation in Mammalian Cells

Several cosmic vectors have been developed to shuttle genes into and out of mammlian cells and to isolate eukaryotic genes by functional selection. These vectors differ from one another chiefly in the prokaryotic and eukaryotic selectable markers they carry and in the restriction sites that can be used for cloning. The type of selectable marker and the presence of autonomous mammalian replicons within the cosmic vector determine the species and types of cells that can be used for transfection.

Cosmid vectors in general have been improved by introducing the following modifications:

1. Two cos sites have been introduced into a single cosmid vector to facilitate the directional cloning.

2. Cosmic vectors have been constructed such that they lack homology with commonly used plasmid vectors. Therefore, these cosmid vectors are unable to recombine with one another if they are carried within the same competent cell. If both the vectors contain homologous foreign DNA sequences, they recombine in recombination competent cells to form cointegrates that can be rescued by transformation. The method may be important for genetic screening to isolate sequences of interest from cosmid libraries.
3. The presence of promoter(s) for bacteriophage encoded DNA dependent RNA polymerases that enables RNA specific to either end of the cloned insert to be generated *in vitro*.
4. Selectable markers have been inserted in the cosmid vectors to simplify the identification of eukaryotic cells that have been transfected with recombinant cosmids.
5. Eukaryotic origin of replication has been inserted in the cosmid vector for making it capable to replicate in mammlian cells after transfection.
6. A range of prokaryotic markers have been inserted to facilitate rescue in bacteria.
7. The use of bacteriophage origin of replication rather than bacterial origin of replication is done. There are some evidences that bacteriophage origins may alleviate the problems of differential growth displayed by recombinant cosmids carrying col E1 origin.

2.9. Shuttle Vectors

If a vector can replicate in different organisms, it is called a shuttle vector. Shuttle vectors contain components like origin of replications, promoters so that these may replicate in different organisms. The first shuttle vector constructed was capable of replicating both in yeast and *E. coli*. Need for such a vector was realized due to the reason that yeast genes were generally not expressed in *E. coli* due to lack of yeast promoters in *E. coli*.

Shuttle vectors contain sequences from *E. coli* plasmid and a particular region of yeast genome. Two origin of replication, one active in *E. coli* and another in yeast, are inserted. Similarly, two selectable markers, Trp detectable in yeast and Amp^r in *E. coli*, are inserted. The genetic map of a shuttle vector as line diagram is shown in Figure 2.12. The gene of interest cloned in it can also be detected in yeast too by virtue of its expression in yeast. Problem with these vectors is that these are unstable in yeast due to lack of centromere. Replicas are not efficiently segregated into daughter cells. To avoid the problem, the recombinant vectors are isolated from a culture obtained from a colony of successfully transformed yeast and then is used to transform *E. coli* where its presence can be detected by ampicillin resistance.

Only one natural yeast plasmid is known which has been used for construction of a series of yeast vectors capable of using yeast as host cell. For the transformation of yeast cells some special procedures are to be used compared to that used for bacterial transformation. The yeast cell wall made up of a polysaccharide, creates hindrance during uptake of the foreign DNA. Therefore, first the cell wall is ruptured using enzymes to form spheroplasts which are made competent by treatment with calcium chloride. After transformation, again the cell wall can be regenerated by culture in a suitable medium. The following are main yeast vectors:

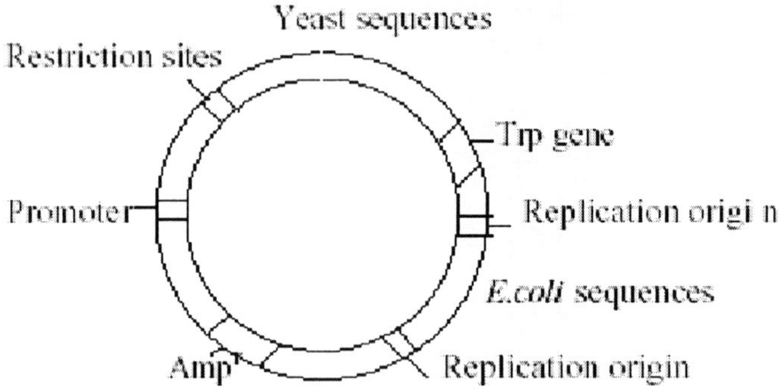

Figure 2.12. Genetic map of a shuttle vector

Yeast Episomal Plasmid Vectors

Certain yeasts contain a 6318 bp long plasmid called as 2μm plasmid. The name is as such due to size of the plasmid. This plasmid carries the origin of replication, a cis acting region (REP3) and two genes REP1 and REP2 (responsible for replication) which together are able to maintain 50 copies of the plasmid in yeast cell stably. The 2μm plasmid has been used for constructing the vectors. In the yeast episomal plasmid vector (Yep), about 50% portion having replication origin and genes required to maintain high copy number of 2μm plasmid is combined with a part of pBR322 to contruct shuttle vector. The portion from pBR322 carries origin of replication of *E.coli* and Amp^r gene as the selectable marker gene for *E.coli* cells. It also carries a yeast gene, HIS 3, acting as marker gene in yeast which encodes an enzyme involved in histidine biosynthesis or LEU2 gene involved in leucine biosynthesis. The naturally occurring promoter of HIS3 gene works both in *E.coli* and in yeast. The genetic map of yeast episomal plasmid vector (YEPC) as line diagram is shown in Figure 2.13.

Yeast Automously Replicating Sequences (ARS)

Origin of replication in yeast is called ARS which is about 100 nucleotides sequence containing a 11 nucleotides consensus sequence necessary for its function. Any double stranded DNA molecule having ARS may replicate inside yeast cells and can be used as yeast vector. Such a DNA molecule is maintained in the yeast cell only if it is a source of some important gene required for survival. For example, TRP1 gene responsible for tryptophan biosynthesis carrying ARS vector, if introduced in mutant yeast cells unable to synthesize tryptophan, will propagate since now TRP1 gene will be available for the survival of the cells.

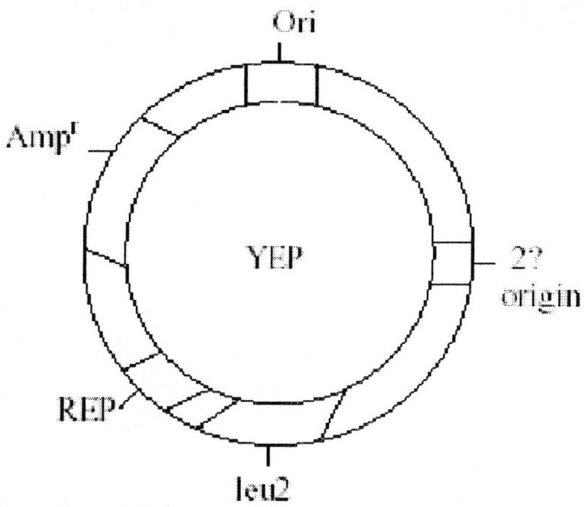

Segments from pBR322 carrying the Ori (replication module), ampicillin resistance gene (Ampr); genes from yeast genome like a selectable marker gene for leucine biosynthesis; region from 2? plasmid of yeast; REP gene responsible for replication.

Figure 2.13. Genetic map of Yeast Episomal plasmid vector

Mini Chromosome Vectors

These vectors behave like very small chromosomes and replicate once during each cell division of the yeast cell and are transferred to the daughter yeast cells like true chromosomes. These vectors contain ARS, a CEN sequence having role in regulation of replication and distribution of the chromosome during cell division, a functional gene like LEU2 which is used as marker gene. Besides, yeast sequences, these also carry *Ori* and Amp^r gene from bacterial plasmid (pBR322). The CEN gene is about 500 bp long obtained from the centromere of yeast chromosomes and is essential for carrying out the centromeric functions of the mini chromosome vector.

Yeast Artificial Chromosomes (YACs)

Telomere, centromere and the replication origin of the yeast chromosomes are isolated and combined to form a DNA capable to propagate like yeast chromosomes. The structure of yeast artificial chromosome as line diagram is shown in Figure 2.14. The two ends of the fragments are ligated to the target DNA to yield a complete chromosome and is transfected in yeast cells. The size of the insert may be more than one Mb, hence may be used for cloning intact eukaryotic genes. For example, the human gene for cystic fibrosis sized 250 Kb is cloned in a YAC vector. These vectors have also been used in cloning human genes and also for mapping of complex eukaryotic chromosomes.

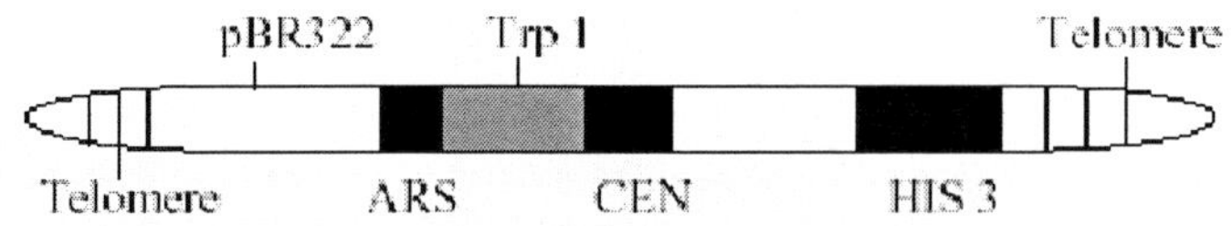

Telomere, centromeric region (CEN), autonomously replication region (ARS)- regions from a yeast chromosome are combined together with region from plasmid pBR322 and a selectable marker gene like HIS (gene for histidine biosynthesis) to form a YAC.

Figure 2.14. Structure of yeast artificial chromosome

The YAC vectors contain segments (ARS, CEN, telomeric segment at both the ends for protecting against exonucleases, selection markers like HIS3 and LEU2) from yeast and also sequences from *E.coli* plasmids. The genetic map of a plasmid based on YAC as line diagram is shown in Figure 2.15. The telomeric sequence contains 20-70 tandem repeats of a 6 bp sequence, 5'CCCCAA3' whose complementary sequence is present on the other strand. Hairpin loop formation occurs at the terminus of the YAC which makes the DNA duplex resistant against exonucleases.

A common YAC vector, pYAC3 contains Tel (telomere region which is extended by telomerase inside the yeast cells), CEN 4 (centromere of chromosome 4 of yeast), ARS, two markers (TRP1 and URA3), an another marker for insertional inactivation that forms the basis of red white screening. The markers enable the yeast cells to grow on a selective medium lacking some nutrients. For example, $TRP1^-$ auxotropic mutants can not grow in a medium lacking tryptophan. If these mutant cells are used as host for the transformation of pYAC 3 vector, only transformed cells will grow on the medium lacking tryptophan as pYAC3 vector contains TRP1 gene which will enable the transformed yeast cells to synthesize tryptophan.

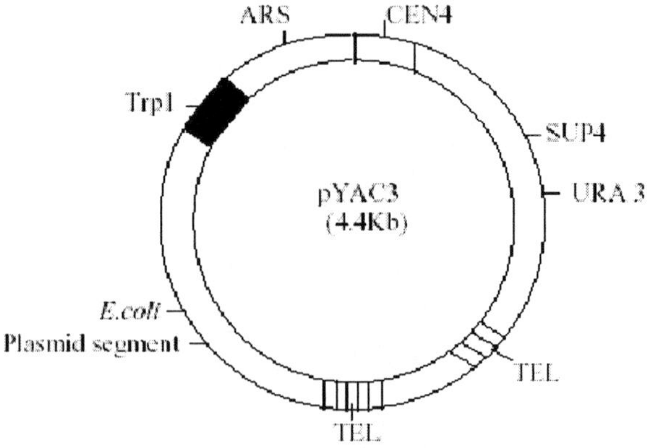

Sequences from yeast chromosome like TEL (telomeres), CEN4 (centomere of yeast chromosome 4), ARS (autonomously replicating sequence) which is replication origin, Trp1 and URA3 selectable markers, SUP4 gene responsible for red white screening are combined with pBR322 sequence.

Figure 2.15. Genetic map of a plasmid based on yeast artificial

2.10. Animal Virus Vectors

Animal virus vectors are also shuttle vectors. Here, use of animal viruses has been taken for constructing the vector since animal viruses go through a large number of rounds of replication when infecting an animal cell. This property has been used to develop vectors commonly called mammalian expression vectors. These vectors are capable of expressing eukaryotic proteins in large quantity. The basic vector (without any insert) is usually grown in *E. coli*, because large amount of the vector can be isolated easily. An example of mammalian expression vector is pgD-DHFR constructed for the expression of glycoprotein D, a protein of herpes simplex virus. Glycoprotein D is important because it has antigenic sites in common with cells infected with various herpes viruses. It is considered that inoculation of human beings with this protein will protect them from subsequent herpes infections. If it is successful, large quantity of glycoprotein D would be needed which may be provided by cultured cells transformed with a vector containing glycoprotein D (gD) gene. PgD-DHFR has been constructed by inserting the followings:

1. A portion of pBR322 containing origin of replication and Amp^r gene has been inserted. The portion enables the vector to be maintained and propagated in *E. coli.*
2. Origin of replication of SV40 tumor virus has been inserted to enable the vector to replicate in mammalian cell lines.
3. A cDNA of mouse dihydrofolate reductase gene (DHFR) has been inserted which acts as selectable marker in mammalian cell lines. In the presence of DHFR gene, the mammalian cell lines are able to grow in a medium lacking hypoxanthine, glycine and thymidine. Otherwise, these components are required in the medium for growth of the mammalian cell lines.
4. The SV40 early promoter has been inserted upstream of restriction sites used for cloning the DNA fragment. This promoter contains TATA box and the mRNA capping site.

Restriction maps of herpes simplex virus DNA were prepared using various restriction endonucleases. A restriction enzyme was selected to cut surrounding the gD gene and generated a fragment containing a stop codon, a transcription stop sequence and a polyadenylation site. Herpes simplex DNA was cleaved with Bam HI, electrophoresed in agarose and the fragment containing gD gene was isolated from the gel. The pgD-DHFR was also cut with Bam HI and ligated with the purified herpes fragment. A mutant strain of Chinese hamster ovary cells which lacked the DHFR gene was transformed with the vector and $DHFR^+$ colonies were isolated by growth in selective medium.

Functional Components of Mammalian Expression Vectors

As written above, mammalian expression vectors are shuttle vectors having prokaryotic sequences that facilitate the propagation of the vector in *E. coli* and eukaryotic transcriptional (expression) units that are expressed in mammalian cells. The mammalian expression unit consists of non-coding sequences and sequences coding for selectable markers derived from different, well characterized viral or mammalian genes.

Prokaryotic sequences that facilitate the propagation in *E. coli* include a replicon capable of functioning in *E. coli*, a gene encoding antibiotic resistance for selection of the recombinants in *E. coli*. It also includes limited number of unique restriction endonuclease sites in non-essential regions to allow insertion of eukaryotic sequences. Most of the vectors contain prokaryotic sequences derived from the plasmid pBR322 or its derivative that lack sequences that are considered to interfere with the replication of the transfected DNA in eukaryotic cells. Deletion of unnecessary segments of plasmid DNA also reduces the size of the vector and also facilitates the insertion of unique restriction sites.

Eukaryotic expression unit contains a promoter element for the transcription of foreign DNA sequences and signals required for efficient polyadenylation of the transcript. Additional elements are enhancers and introns with functional splice donor and acceptor sites. Promoters and enhancers consist of short DNA sequences that interact specifically with cellular proteins involved in transcription. The combination of different recognition sequences and the amounts of the cognate transcription factors determine the efficiency for transcription of a gene in a particular cell line.

Mostly, eukaryotic promoters contain TATA box and the upstream promoter elements. The TATA box located 25-30 bp upstream from the transcription initiation site is considered to be involved in directing RNA polymerase II to start RNA synthesis correctly. Upstream promoter elements control the rate of the transcription. These elements may act in either orientation and are located 100 to 200 bp upstream of the TATA box.

Enhancer sequences downstream of the transcription initiation site and present at considerable distances from the promoter may stimulate transcription upto 1000 fold. Some enhancers become active only in the presence of an inducer such as a hormone or a metal ion etc. Many enhancer elements derived from viruses have a broader host range and are active in a variety of tissues. SV40 early gene enhancer is active in many mammalian cell lines, therefore, the vectors having this enhancer have been widely used.

RNA polymerase II transcribes through the site where polyadenylation occurs. Subsequently, 3' terminus of mRNA is formed by site specific post transcriptional cleavage and polyadenylation. Generally, termination of the primary transcript occurs a few hundred nucleotides downstream from the polyadenylation site. Two sequences, GU or U rich sequences located downstream from the polyadenylation site, and highly conserved AAUAAA sequence located 11-30 nucleotides upstream, are required for accurate and efficient polyadenylation. Therefore, sequences downstream from the polyadenylation site must be incorporated in eukaryotic expression vectors. A 237 bp Bam HI- Bcl I restriction fragment of SV40 containing the cleavage and polyadenylation signals from both the early and the late transcription unit has been mostly incorporated in the mammalian expression vectors. These signals are positioned in opposite orientations, one on each strand and both sets of signals are shown to be extremely efficient for the processing of hybrid mRNAs. These 3' non-coding sequences are also considered to play role in mRNA stability.

In eukaryotes, DNA sequences coding for an eukaryotic protein are separated in the genome by intervening sequences (introns) that may vary in size and number. An intron may be from tens to many thousands of nucleotides. After polyadenylation, introns are removed by splicing to make mRNA which is transported from the nucleus to the cytoplasm. The sequences required for splicing are found at the 5' and 3' boundaries of the intron and are much shorter in size compared to the size of the intron.

Phagmid Vectors

The vectors which are constructed by combination of plasmids and phage sequences are called phagmid vectors. pKUN 9 and pKUN 19 are the examples of phagmid vectors. Besides, there are number of other phagmid vectors. The vectors, pKUN 9 and pKUN 19 are derivatives of corresponding pUC 9 and pUC 19 vectors. These vectors have the following properties:

1. The colE1 origin of replication has been inserted which enables the vector to replicate as a high copy number vector in the absence of helper phage.
2. The *bla* gene encoding for β-lactamase has been inserted which due to ampicillin resistant character is used as a selectable marker for transformation.
3. A fragment of *E. coli lac* operon having regulatory region and a short fragment encoding the first 77 amino acids (α-peptide) of β-galactosidase and containing multiple cloning sites linker has been inserted. The sequence helps in carrying α-complementation. Details about α-complementation are given above under pUC plasmids.
4. A fragment containing the morphogenetic signal and the viral (+) and complementary (-) strand replication origins of Ff (nucleotides from 5488 to 6001 of f1) has been cloned in the Nar I site of the vectors.
5. A fragment containing the morphogenetic signal and the viral (+) and complementary (-) strand replication origins of Ike (from nucleotides 5921 to 6621) has been inserted in the unique Nde I site in an orientation opposite to that of Ff.

2.11. Ti Plasmids

Agrobacterium tumefaciens is a gram negative soil bacteria which produces crown gall in dicotyledonous plants and can mediate transfer of genetic material into plants. *Agrobacterium* has a plasmid in it called as Ti plasmid. The genetic map of a Ti plasmid as line diagram is shown in Figure 2.16. The Ti plasmid has been modified to use these as gene transferring agents in plants. The tumor formation after the invasion of the bacteria occurs due to expression of genes on the T-DNA that stimulates the tumor formation. The property of tumor formation can be eliminated by disarming the strain through deletion of genes in the T-DNA without loss of DNA transfer and integration functions. For genetic transformation, the gene to be transferred is cloned in T-DNA and is linked to a gene conferring resistance to a selectable marker (like antibiotic resistance, Kan^r). Only cells having manipulated T-DNA will show cell division and development in the presence of the selective marker.Leaf disc, stem fragments, protoplasts have been used for transferring cloned T-DNA to recover genetic transformants. Using Ti vectors, genes for resistance to herbicides, insecticides, pests have been successfully transferred to economically important crops, vegetables and ornamental plants.

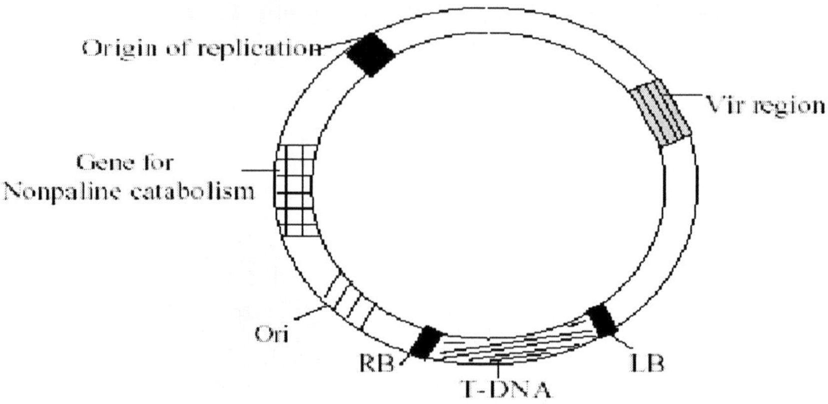

Vir- virulence region; Ori- origin of replication in *Agrobacterium*; RB- right border sequence; LB- left border sequence

Figure 2.16. Genetic map of the Ti plasmid

Development of Ti plasmids depends on the two properties of DNA that is transferred from *Agrobacterium* and stably inserted into the plant genome. For construction of Ti plasmid, first T-DNA is bounded by two directly repeating sequences of 25 bp called as border sequences, essential for transformation. The line diagram of T-DNA is shown in Figure 2.17 and nucleotide sequences of the border sequences are shown in Figure 2.18. None of the genes present in T-DNA between the border sequences are essential for DNA transfer. All the DNA sequences between border sequences which also have phytohormone synthesizing genes and crown gall forming genes are deleted. This type of plasmid is also called as Disarmed Ti plasmid. A plasmid, pTiB6S3-SE has been constructed which retains the left border sequence and a portion of an octopine type Ti plasmid T-DNA. The genetic map of pTiB6S3 as line diagram is shown in Figure 2.19. This segment of T-DNA provides homology for cointegrate formation. Other T-DNA segments or heterologous DNAs such as pBR322 sequences may also be used to provide homology.

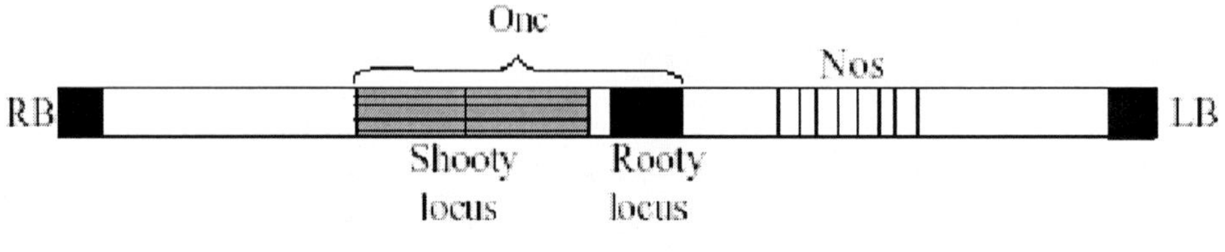

RB-right border sequence; LB- left border sequence; Onc -oncogenes; Nos-nopaline catabolism gene

Figure 2.17. Diagram representing the region of T-DNA of the Ti plasmid

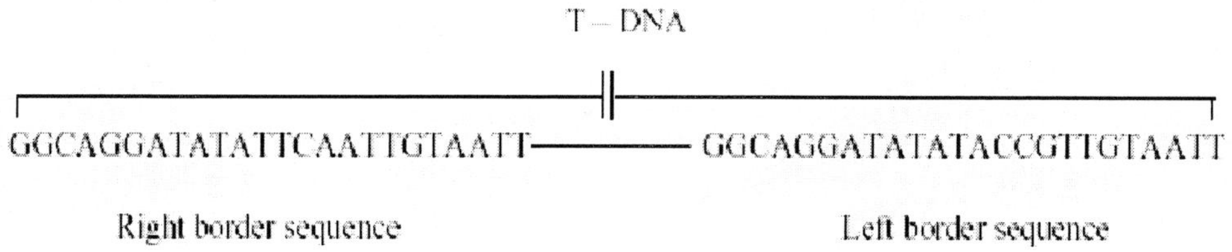

Figure 2.18. Sequence of the right and left borders of the T-DNA

The cointegrating vectors also require intermediate plasmids that are used for cloning and analysis in *E. coli* and are subsequently introduced into *Agrobacterium* cells where they combine with the Ti plasmid to give a cointegrate that contains hybrid T-DNA. A plasmid named as pMON200 has been constructed for the purpose. It has 1.6 Kb left inside homology (LIH) segment derived from the octopine type pTiA6 plasmid. The segment is responsible for help in recombination with the resident Ti plasmid in *Agrobacterium.* The plasmid also contains a 1.6 Kb fragment carrying origin of replication of pBR322, a 2.2 Kb fragment of nopaline type pTiT37 that carries the right border sequences of nopaline T-DNA (NRB) and intact nopaline synthase gene (NOS), a 2.7 Kb fragment of Tn7 carrying spectinomycin/streptomycin resistant genes (Spc/Str^r), a 1.6 Kb fragment having chimeric NOS-NPTII'-NOS gene capable of acting as selectable marker (kanamycin resistance), and a synthetic multicloning site for insertion of DNA fragments. The genetic map of pMON200 as line diagram is shown in Figure 2.20.

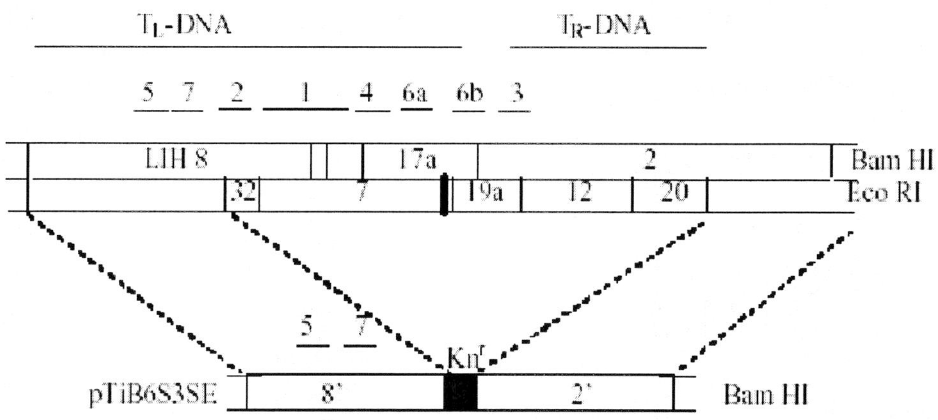

Figure 2.19. Construction and structure of the disarmed pTiB6S3-SE plasmid. A 4.5-Kb Bam HI-EcoRI fragment 8' (from Bam HI fragment 8) and a 6.0-Kb Eco RI-Bam HI fragment 2' (from Bam HI freagment 2) provided homology for replacement of most of greater than 80% of the TL T-DNA and the entire TR DNA with the 1.2 Kb Tn 903 segment carrying bacterial kanamycin resistance. The resulting pTiB6S3-SE plasmid carries only the TL left border and intact transcripts 5 and 7 which are not involved in disease production.

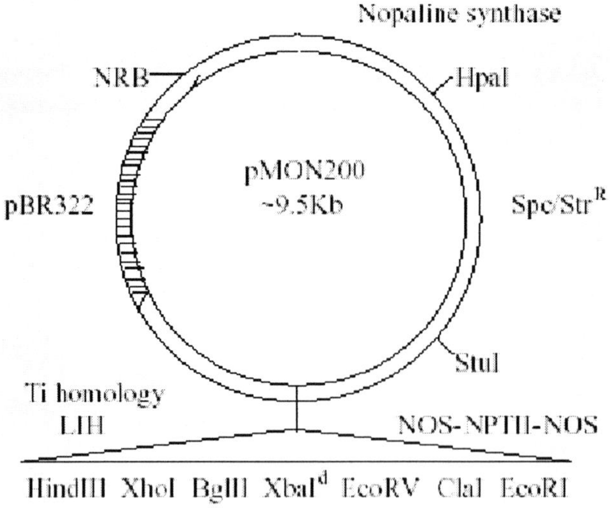

Figure 2.20. Genetic Map of pMON200. All of the restriction endonuclease sites shown are single. The 'd' superscript denotes a site that is cleavable only in DNA prepared from a *dam* methylase deficient *E.coli* host. NRB- right border sequence of nopaline T- DNA, NOS- nopaline synthase gene.

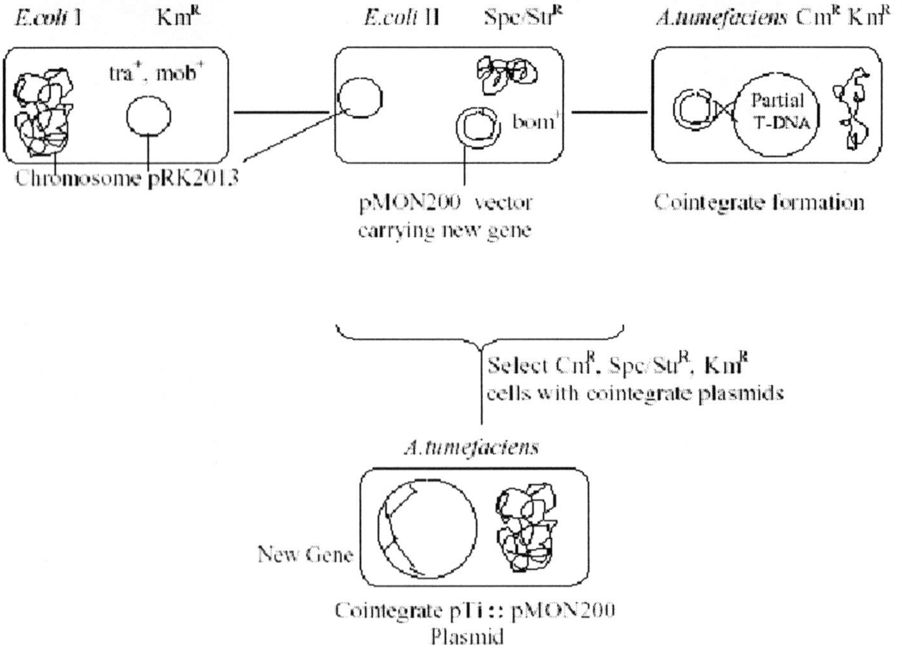

Figure 2.21.Tri-parental mating procedure using pMON200 into *Agbro-bacterium tumefaciens* to produce a cointegrate plasmid carrying a hybrid T-DNA.

DNA fragment to be transferred in a plant cell is first inserted in the plasmid, pMON200 using its unique restriction site present within the multicloning region. The recombinant, pMON200 is transformed in *E. coli* cell. *E. coli* cells carrying pRK2013, a helper plasmid, are mixed with *E. coli* cells containing pMON200 and *Agrobacterium tumefaciens* cells having pTiB6S3-SE helper plasmid are added. The pRK2013 plasmid provides RK2 transfer functions and ColE1 *mob* protein that act at the pMON200 *bom* site to mobilize pMON200 into *Agrobacterium*. The plasmid, pMON200 can not replicate in *Agrobacterium*, therefore, it recombines with pTiB6S3-SE helper plasmid for its maintenance. The *Agrobacterium* cells also become spectinomycin and streptomycin resistant after pTiB6S3-SE and pMON200 cointegration in addition to already chloramphenicol and kanamycin resistant. Transformation of a DNA in *Agrobacterium* through pMON200 and pRK2013 using *E. coli* as intermediary host is shown in Figure 2.21. When *Agrobacterium* cells having cointegrate are co-cultivated with plant cells or explants, the hybrid T-DNA containing any new gene located between the border sequences is transferred in plant cell and becomes a part of the genome of the plant cell. The process is called mating. That's way, with the help of Ti plasmid, one can transform a gene in the plant cells.

2.12. Expression Vectors

Vectors having strong promoters which are required for efficient transcription are capable of expressing the cloned genes. Such vectors are called as expression vectors. In many cases, fused promoters (sequences of more than one promoter are fused to make a strong promoter) have been used and construction is done in such a way that the DNA fragment is cloned within a gene which gets expressed. In many cases, *lac* Z gene is used for the purpose and after cloning, the cloned gene is expressed as a fusion protein with β-galactosidase protein. Here, we describe few expression vectors in use:

pUR Vectors

Ruther and Muller-Hill constructed a series of vectors called as pUR vectors. First, they constructed a series, pUR278, pUR288 and pUR289 having three different open reading frames. If a gene is cloned in all the three vectors using the same restriction site, in one of these three, it will be in a proper reading frame. These vectors have multiple cloning site having recognition sequences of Bam HI, Sal I, Xba I, Hind III and Cla I at the 3'-terminus of the *lac* Z gene. Pst I site is present within the ampicillin resistant gene. The size of the vectors is nearly 5.2 Kb. The genetic map of pUR278 as line diagram is shown in Figure 2.22. Afterwards, they constructed a new series namely pUR290, pUR291 and pUR292 where Pst I site has been inserted within multiple cloning site region after destroying Pst I site in the Amp^r gene. The sequences of the multiple cloning site regions of these vectors are shown in the Figure 2.23.

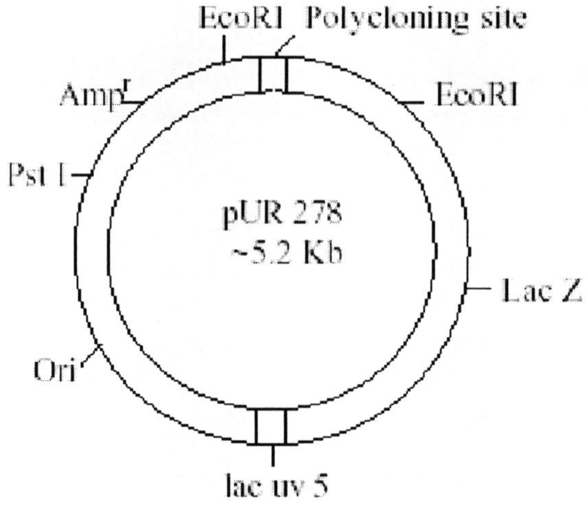

Figure 2.22. Genetic map of pUR 278

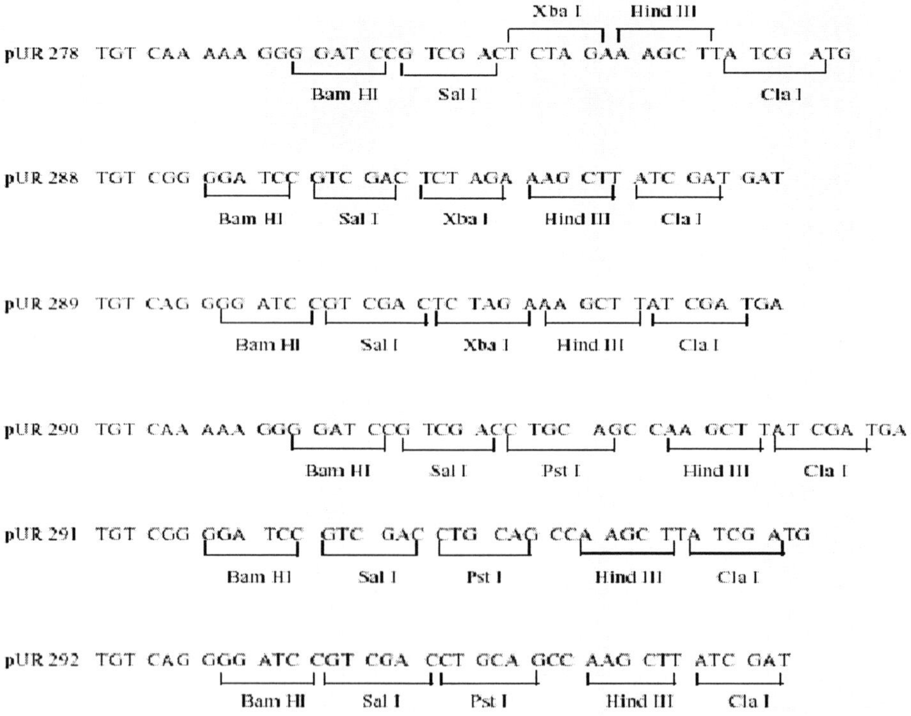

Figure 2.23. Sequences of the multiple cloning sites in pUR vectors

pEX Vectors

Stanley and Luzio constructed pEX vectors series namely pEX1, pEX2 and pEX3 in three different reading frames. In the vector, λ P_R promoter has been inserted which directs the expression of Cro-β-galactosidase fusion protein. Polycloning site containing recognition sequences of EcoRI, Sma I, Bam HI, Sal I and Pst I is located in the 3' end region of *lac* Z gene. The genetic map of pEX3 as line diagram is shown in Figure 2.24. Translation and transcription stop signals are present downstream of the polycloning site. It contains Amp^r gene as selective marker and origin of replication of pBR322. The amino terminus part of the *lac* Z gene has been substituted by some sequences of bacteriophage λ Cro gene and *lac* I gene of *E. coli.* It also has bacteriophage λ cIts 857 repressor which regulates the expression of the fusion protein.

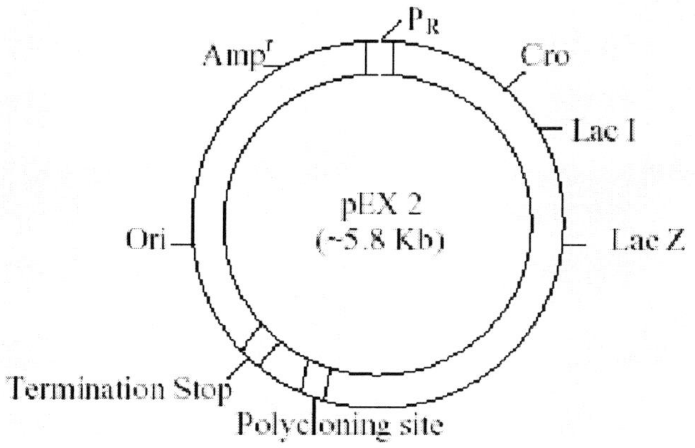

PR- ? PR promoter; Cro – portion of Cro gene fused with Lac Z (? -galactosidase) gene; Lac I- portion of Lac I gene fused with Lac Z gene; Ori- origin of replication.

Figure 2.24. Genetic Map of pEX 2

pMR100 Vector

The vector has been developed by Gray et. al. It is an open reading frame (ORF) vector. Its size is nearly 11.6 Kb. It has been designed in such a way that *lac* Z is not in frame with sequences encoding the amino terminus of the protein and therefore, there is very low expression of β-galactosidase. Insertion of a DNA fragment that has an open reading frame and that places the *lac* Z sequences in frame results in a vector expressing a protein that has β-galactosidase activity. After cDNA fragments are inserted into the ORF vector, *E. coli* strain LG 90 which has deletion of *lac* Z, is transformed and plated on MacConkey lactose indicator plates. Cells expressing high levels of β-galactosidase are screened for the protein synthesized. It carries a *lac* promoter which directs the transcription of part of the cI gene of bacteriophage λ, followed by a polycloning site and an out of frame *lac* Z gene deleted at its 5'-terminus. The genetic map of pMR100 as line diagram is shown in Figure 2.25.

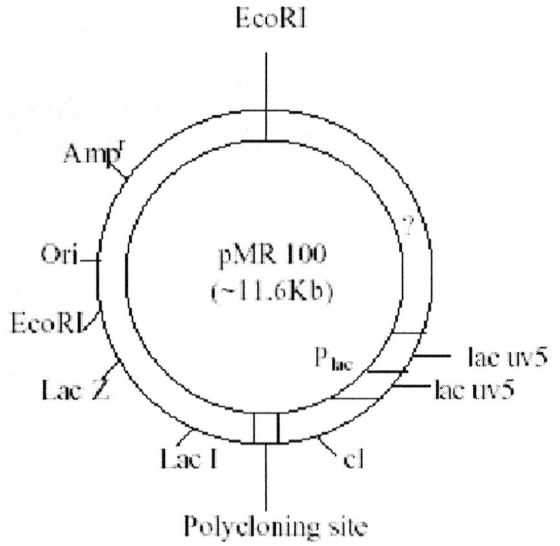

Figure 2.25. Genetic map of pMR100. Plac is a double lac uv5 promoter

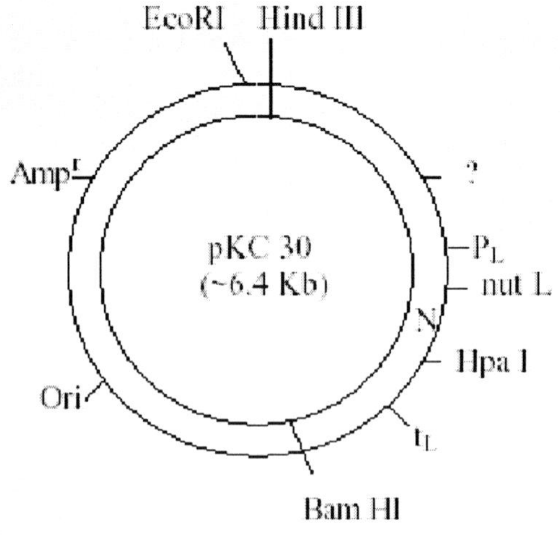

Figure 2.26. Genetic map of pKC 30

pKC 30 Vector

The vector has been developed using P_L promoter of bacteriophage λ and a Hpa I recognition site located 321 nucleotides downstream from the P_L transcriptional start site. The size of the vector is nearly 6.4 Kb. The genetic map of pKC30 as line diagram is shown in Figure 2.26. It is a derivative of pBR322 and contains a Hind III- Bam H I fragment derived from bacteriophage λ inserted between the Hind III and Bam H I sites within the tet^r gene. The insertion contains the promoter signal P_L, a site recognized by the N gene product (Nut L), N gene itself and the strong rho-dependent transcription termination signal t_L. The Hpa I recognition site lies within the coding region of the N gene. Sequences inserted into the Hpa I site can be regulated by introducing the recombinant plasmid into a temperature sensitive bacteriophage λ lysogen (cIts 857). The cells may be grown to mid log phase at 30°C and then shifted to 40°C to inactivate the cI gene product and to turn on the P_L promoter.

References

[1] Sambrook, J., Fritsch, E.F. and Maniatis, T. (1989) *Molecular Cloning* Vol. 1, Cold Spring Harbor Laboratory Press, New York.

[2] Bolivar, F., Rodriguez, R.L., Betlach, M.C. and Boyer, H.W. (1977) Construction and characterization of new cloning vehicles: I- Ampicillin resistant derivatives of the plasmid oMB9. *Gene* **2**: 75-93.

[3] Bolivar, F., Rodriguez, R.L., Greene, P.J., Betlach, M.C., Heyneker, H.L., Boyer, H.W., Crosa, J.H. and Felkow, S. (1977) Construction and characterization of new cloning vehicles: II- A multipurpose cloning system. *Gene* **2**: 95-113.

[4] Messing, J. (1983) New M13 vectors for cloning.*Methods Enzymol.* **101**: 20-78.

[5] Vieira, J. and Messing, J. (1982) The pUC plasmids, an M13mp7-derived system for insertion mutagenesis and sequencing with synthetic universal primers. *Gene* **19**: 259-268.

[6] Glover, D.M.. (1988) *DNA Cloning Techniques: A Practical Approach.* Vol. 1, IRL Press, Oxford.

[7] *M13 Cloning and Sequencing handbook*, A booklet of Amersham.

[8] Perron, Y.C., Vieira, J. and Messing, J. (1985) Improved M13 phage cloning vectors and host strains: Nucleotide sequences of the M13mp18 and pUC19 vectors. *Gene* **33**: 103-119.

[9] Heuermann, K. and Cosgrove, J. (2001) Life Science: A Technical Application *Newsletter* 2 (2) 2-5.

Chapter III

Restriction Endonucleases

3.1. Introduction

Restriction endonucleases bind specifically to and cleave the sugar phosphate backbone (bond between deoxyribose and phosphate groups) in double stranded DNA at specific sites within or adjacent to a particular sequence known as recognition sequence. Restriction endonucleases are of three types: Type I, Type II and Type III. Types I and III restriction endonucleases carry modification (like methylation) and ATP dependent cleavage activities in the same protein. Type III restriction endonucleases cleave the DNA at the recognition site and then dissociate from the substrate. However, type I restriction endonucleases bind to the recognition sequence but cleave at random sites when the DNA loops back to the bound enzyme. Neither type I nor type III restriction endonucleases are widely used in genetic engineering. Type II restriction endonucleases are commonly used in genetic engineering. They are the invaluable molecular scissors. These restriction endonucleases cut both strands of duplex DNA with in a stretch of just a few bases. Therefore, only type II restriction endonucleases are described in this chapter. Nomenclature of the restriction endonucleases is based on the source of their isolation from an organism. In case of restriction endonucleases, enzyme activity is described in units considering one unit of the restriction endonuclease as the amount of the enzyme required to hydrolyze one microgram of lambda DNA (λ DNA) to completion in one hour at the optimum temperature in a total volume of 50 μl of the assay mixture. The lambda DNA, due to its large size (approximately 50 Kb), is considered to have recognition sites almost for all the restriction endonucleases.

Factors like pH, salt concentration, methylation of bases within the recognition sequence affect the activity of the restriction endonuclease. pH and salt optima for restriction endonucleases are generally broad. Generally, upon methylation of a base within the recognition sequence, recognition sequence no more remains recognition sequence for the particular restriction endonuclease. However, certain restriction endonucleases still recognize the sequence, such type of restriction endonucleases are said to have ‘Star’ activity. Star activity is a change or relaxation of specificity. It has been observed under specific conditions that affect hydrogen bonding like high enzyme concentrations with deviation from recommended enzyme assay and reaction buffers (such as substitution of other metal ions for

Mg^{+2}, conditions of high pH or low salt concentrations) or the use of various organic solvents.

Nowadays, nearly 1000 different restriction endonucleases are known and most of them are also available commercially, some of the important and commonly used restriction endonucleases are described below. If more than one restriction endonucleases isolated from different organisms recognize the same recognition sequence, they are called 'Isoschizomers'.

All these restriction endonucleases have been isolated from bacteria where they are considered to serve defense purpose. When a foreign DNA from an infecting virus enters into a bacteria, the restriction endonuclease present inside the bacteria will cleave foreign DNA to inactivate the infecting virus. These restriction endonucleases do not chew the genomic DNA with in the bacteria. In most cases, bacteria that synthesizes a particular restriction endonuclease also synthesizes a companion DNA methyl transferase. The DNA methyl transferase methylates either adenine or cytosine of DNA recognition sequence for that restriction endonuclease and methylase is sometimes called as restriction-modification system.

Conventionally, restriction endonucleases are named after their host of origin. For example, Bam H I was isolated from *Bacillus amyloliquefaciens* strain H. The first letter of the enzyme indicates genus and another two letters indicate species. In some cases, strain name letter is also included in the name of the enzyme. If from the same bacteria, more than one restriction endonuclease are isolated, then I and II are used in naming the enzyme.

It was considered earlier that three different buffers called as low salt, medium salt and high salt buffers are sufficient for all the restriction endonucleases to perform their optimal activity but nowadays, for every restriction endonuclease, specific optimum buffer is described. Generally, commercial companies which market these enzymes also supply the buffer (10x) for that particular restriction endonuclease. The following are the three buffers commonly used for restriction endonucleases:

Buffer	NaCl	Tris, pH 7.5	$MgCl_2$	DTT
Low salt	0 mM	10 mM	10 mM	1 mM
Medium salt	0 mM	10 mM	10 mM	1 mM
High salt	100 mM	50 mM	10 mM	1 mM

The restriction endonucleases cleave at specific site with in a specific sequence of double stranded DNA called as recognition sequence. The length of the recognition sequence varies, some restriction endonucleases recognize 6 base pairs (bp) sequence of DNA, whereas others recognize a sequence of 8 bp. Some restriction endonucleases have recognition sequence of only 4 bp. The length of the recognition sequence dictates how frequently the enzyme will cleave in a random sequence of DNA. It is considered that a restriction endonuclease with a 6 bp recognition sequence cuts on an average every 4096 bp, whereas, a restriction endonuclease with a 4 bp recognition sequence cleaves roughly every 256 bp. The shorter the recognition sequence length, more is the frequency of its occurrence and *vice –versa*.

Each recognition sequence of restriction endonucleases has been found to be palindromic (they read identical when read in each strand from 5' to 3'). Most restriction endonucleases bind to their recognition sequence as dimers. The recognition sequence may be ambiguous or unambiguous. Some restriction endonucleases have well defined nucleotide sequences and will not recognize if there is any change in it. Whereas, many restriction endonucleases may have variability in recognition sequence. You will find below that in many cases, with in recognition sequence, 'N' is written which stands for any nucleotide. In some cases, 'py' or 'pu' is written where 'py' stands for pyrimidine and 'pu' stands for purine nucleotides.

As written above, restriction endonucleases cleave only double stranded DNA by recognition of specific sequences with in the DNA molecule. However, some restriction endonucleases cleave the DNA at some distance from the recognition sequence.

All DNA molecules cleaved with the same restriction endonuclease have identical termini regardless of the source of the DNA. The use of this property has been made in DNA recombination studies where DNA molecules from evolutionary distinct organisms may be ligated to create recombinant DNA.

Upon restriction endonuclease digestion, DNA may generate blunt ends, 5' overhang (protruding, cohesive) ends or 3' overhang ends depending on the position of the cutting site within the recognition sequence. The 5'- overhang ends occur when the enzyme cuts to the left of the central axis of symmetry, whereas, 3'- protruding ends occur when the enzyme cuts to the right of the central axis of symmetry. The blunt ends occur when the enzyme cuts directly at the central axis of symmetry. Protruding ends are also called as 'sticky' or 'staggered' ends since the central axis of a recognition sequence are complementary. These ends will anneal or stick to one another if brought together in close proximity. DNA molecules with protruding ends are more easily annealed in a ligation reaction compared to DNA molecules having blunt ends.

3.2. Important Restriction Endonucleases Frequently Used In

Genetic Engineering

i. Acc I

The restriction endonuclease, Acc I has been isolated from *Acinetobacter calcoaceticus* and the recognition sequence for this enzyme is given below. The arrows indicate the cutting site within the sequence. As per international norms, the sequence is written from 5' to 3' end in the sense strand and 3' to 5' end in the antisense strand:

AG
↓
GT CT AC

CAGA TG
↑
TC

The AG and TC are written above CT and below GA in the sense and the antisense sequence which indicates that if C is replaced by A and/or T by G, still it will be the recognition sequence for the enzyme. The same stands true for the sequences mentioned for other restriction endonucleases, if written in the same fashion. The enzyme produces 5' protruding ends.

The optimum assay buffer for the enzyme activity is 10 mM Tris-HCl, pH 8.0, 10 mM NaCl, 10 mM $MgCl_2$, 1 mM 2-mercaptoethanol (2-ME) and 100 μg bovine serum albumin (BSA)/ml. The optimum storage medium is 10 mM Tris-HCl, pH 7.5, 50 mM KCl, 1 mM dithiothreitol (DTT), 0.1 mM EDTA, 300 μg BSA/ml and 50 % glycerol (v/v). The optimum storage temperature is –20 °C and the optimum assay temperature is 37 °C. The enzyme is fully active up to 5 hours in the assay system. On methylation of A (GTCTmAC), it is no more cut by the enzyme. As mentioned above, there may be variation in the optimum conditions depending upon the source from where the enzyme is purchased. Nowadays, generally manufacturers supply 10x reaction buffer with the enzyme.

ii. Alu I

The restriction endonuclease, Alu I has been isolated from *Arthrobacter luteus*. The recognition sequence and cutting site within the sequence is shown below:

↓
AG CT

TC GA
↑

The enzyme produces blunt ends. The optimum assay buffer for the enzyme activity is 10 mM Tris-HCl, pH 7.5, 50 mM NaCl, 10 mM $MgCl_2$, 10 mM 2-ME and 100 μg BSA/ml. The optimum storage medium for the enzyme is 10 mM Tris-HCl, pH 7.5, 50 mM KCl, 1 mM DTT, 0.1 mM EDTA, 300 μg BSA/ml and 50 % glycerol (v/v). The enzyme is stored at –20 °C. The optimum temperature for the enzyme assay is 37 °C. The enzyme is fully active up to 5 hours in the assay system. If methylation of C (AGmCT) occurs, then the sequence is no longer cut by the enzyme.

iii. Apa I

The restriction endonuclease, Apa I has been isolated from *Acetobacter pasteurianus*. The recognition sequence and the cutting site within the sequence is shown below:

↓
GGGCC C
C CCGGG
↑

The enzyme produces 3' protruding ends. The optimum assay buffer for the enzyme activity is 10 mM Tris-HCl, pH 7.5, 10 mM $MgCl_2$, 10 mM 2-ME and 100 μg BSA/ml. The optimum storage medium for the enzyme is 10 mM Tris-HCl, pH 7.5, 50 mM KCl, 1 mM DTT, 0.1 mM EDTA, 300 μg BSA/ml and 50 % glycerol (v/v). The enzyme is stored at –20 °C. The optimum temperature for the enzyme assay is 37 °C. The enzyme is fully active up to 5 hours in the assay system. If methylation of C (GGGmCCC) occurs, then the sequence is not cut by the enzyme.

iv. Ava I

The restriction endonuclease, Ava I has been isolated from *Anabaena variabilis*. The recognition sequence and the cutting site within the sequence is shown below:

C↓ *A*
CT CGGG
GAGCC↑ C
G T

The enzyme produces 5' protruding ends. The optimum assay buffer for the enzyme activity is 10 mM Tris-HCl, pH 8.0, 60 mM NaCl, 10 mM $MgCl_2$, 1 mM 2-ME and 100 μg BSA/ml. The optimum storage medium for the enzyme is 10 mM Tris-HCl, pH 7.5, 50 mM KCl, 1 mM DTT, 0.1 mM EDTA, 300 μg BSA/ml and 50 % glycerol (v/v). The enzyme is stored at –20 °C. The optimum temperature for the enzyme assay is 37 °C. The enzyme is fully active up to 3 hours and partially active up to 5 hours in the assay system. If methylation of C (CTmCGGG) occurs, then the sequence is not cut by the enzyme. The enzyme may exhibit star activity if the assay buffer is modified.

v. Ava Ii

The restriction endonuclease, Ava II has been isolated from *Anabaena variabilis*. The recognition sequence and the cutting site within the sequence is shown below:

A
G ↓GTCC
CCAG ↑G
T

The enzyme produces 5' protruding ends. The optimum assay buffer for the enzyme activity is 10 mM Tris-HCl, pH 8.0, 60 mM NaCl, 10 mM $MgCl_2$, 1 mM 2-ME and 100 μg BSA/ml. The optimum storage medium for the enzyme is 10 mM Tris-HCl, pH 7.5, 50 mM KCl, 1 mM DTT, 0.1 mM EDTA, 300 μg BSA/ml and 50 % glycerol (v/v). The enzyme is stored at –20 °C. The optimum temperature for the enzyme assay is 37 °C. The enzyme is fully active up to 3 hours and partially active up to 5 hours in the assay system. If methylation of C (GGTCmC) takes place, then the sequence is not cut by the enzyme.

vi. BamH I

The restriction endonuclease, Bam H I has been isolated from *Bacillus amyloliquefaciens* H. The recognition sequence and the cutting site within the sequence is shown below:

G ↓ GATCC

CCTAG ↑ G

The enzyme produces 5' protruding ends. The optimum assay buffer for the enzyme activity is 10 mM Tris-HCl, pH 8.0, 100 mM NaCl, 10 mM $MgCl_2$, 1 mM 2-ME and 100 μg BSA/ml. The optimum storage medium for the enzyme is 10 mM Tris-HCl, pH 7.5, 100 mM KCl, 1 mM DTT, 0.1 mM EDTA, 300 μg BSA/ml and 50 % glycerol (v/v). The enzyme is stored at –20 °C. The optimum temperature for the enzyme assay is 37 °C. The enzyme is fully active up to one hour and partially active up to 2 hours in the assay system. If methylation of C or A (GGATCmC, GGmATCC) is there, the sequence is still cut by the enzyme, whereas, GGATmCC is not cut.

vii. Bcl I

The restriction endonuclease, Bcl I has been isolated from *Bacillus caldolyticus.* The recognition sequence and the cutting site within the sequence is shown below:

T ↓ GATCA

ACTAG ↑ T

The enzyme produces 5' protruding ends. The optimum assay buffer for the enzyme activity is 10 mM Tris-HCl, pH 8.0, 75 mM KCl, 10 mM $MgCl_2$, 1 mM 2-ME and 100 μg BSA/ml. The optimum storage medium for the enzyme is 10 mM Tris-HCl, pH 7.5, 50 mM KCl, 1 mM DTT, 0.1 mM EDTA, 300 μg BSA/ml and 50 % glycerol (v/v). The enzyme is stored at –20 °C. The optimum temperature for the enzyme assay is 50 °C. The enzyme is fully active up to 5 hours in the assay system. If interior methylation of A (TGmATCA, TGATCmA) is there, the sequence is not cut by the enzyme.

viii. Bgl I

The restriction endonuclease, Bgl I has been isolated from *Bacillus globigii* RUB 561. The recognition sequence and the cutting site within the sequence is shown below:

GCCNNNN ↓ NGGC

CGGN ↑ NNNNCCG

The enzyme produces 3' protruding ends. The optimum assay buffer for the enzyme activity is 10 mM Tris-HCl, pH 8.0, 100 mM KCl, 10 mM $MgCl_2$, 10 mM 2-ME and 100 μg BSA/ml. The optimum storage medium for the enzyme is 10 mM Tris-HCl, pH 7.5, 300 mM KCl, 1 mM DTT, 0.1 mM EDTA, 300 μg BSA/ml and 50 % glycerol (v/v). The enzyme is stored at –20 °C. The optimum temperature for the enzyme assay is 37 °C. The enzyme is fully active up to one hour and partially active up to 5 hours in the assay system.

ix. Bgl II

The restriction endonuclease, Bgl II has been isolated from *Bacillus globigii* RUB 562. The recognition sequence and the cutting site within the sequence is shown below:

A ↓ GATCT

TCTAG ↑ A

The enzyme produces 5' protruding ends. The optimum assay buffer for the enzyme activity is 10 mM Tris-HCl, pH 8.0, 100 mM KCl, 10 mM $MgCl_2$, 10 mM 2-ME and 100 μg BSA/ml. The optimum storage medium for the enzyme is 10 mM Tris-HCl, pH 7.5, 300 mM KCl, 1 mM DTT, 0.1 mM EDTA, 300 μg BSA/ml and 50 % glycerol (v/v). The enzyme is stored at –20 °C. The optimum temperature for the enzyme assay is 37 °C. The enzyme is fully active up to one hour and partially active up to 5 hours in the assay system. The enzyme exhibits approximately two fold activity at pH 8.0 as compared to at pH 7.4. If methylation of A (AGmATCT) is there, the sequence is cut by the enzyme. Whereas, if methylation of C (AGATmCT) is there, the sequence is not cut by the enzyme.

x. Bst I

The restriction endonuclease, Bst I has been isolated from *Bacillus stearothermophilus.* The recognition sequence and the cutting site within the sequence is shown below:

G ↓ GATCC

CCTAG ↑ G

The enzyme produces 5' protruding ends. The optimum assay buffer for the enzyme activity is 10 mM Tris-HCl, pH 7.5, 50 mM NaCl, 10 mM $MgCl_2$, 1 mM 2-ME and 100 μg BSA/ml. The optimum storage medium for the enzyme is 10 mM Tris-HCl, pH 7.5, 200 mM KCl, 1 mM DTT, 0.1 mM EDTA, 300 μg BSA/ml and 50 % glycerol (v/v). The enzyme is stored at –20 °C. The optimum temperature for the enzyme assay is 55 °C. The enzyme assay is generally carried out under the paraffin oil to avoid evaporation. The enzyme is fully active up to one hour and partially active up to 5 hours in the assay system. Under certain conditions, some times the enzyme exhibits star activity.

xi. BstE II

The restriction endonuclease, BstE II has been isolated from *Bacillus stearothermophilus* ET. The recognition sequence and the cutting site within the sequence is shown below:

G ↓ GTNACC
CCANTG ↑ G

The enzyme produces 5' protruding ends. The optimum assay buffer for the enzyme activity is 10 mM Tris-HCl, pH 8.0, 150 mM NaCl, 10 mM $MgCl_2$, 10 mM 2-ME and 100 μg BSA/ml. The optimum storage medium for the enzyme is 10 mM Tris-HCl, pH 7.5, 50 mM KCl, 1 mM DTT, 0.1 mM EDTA, 300 μg BSA/ml and 50 % glycerol (v/v). The enzyme is stored at –20 °C. The optimum temperature for the enzyme assay is 60 °C. The enzyme assay is generally carried out under the paraffin oil to avoid evaporation. The enzyme is fully active up to one hour and partially active up to 5 hours in the assay system.

xii. Cla I

The restriction endonuclease, Cla I has been isolated from *Caryophanon latum*. The recognition sequence and the cutting site within the sequence is shown below:

AT ↓ CGAT
TAGC ↑ TA

The enzyme produces 5' protruding ends. The optimum assay buffer for the enzyme activity is 10 mM Tris-HCl, pH 8.0, 50 mM KCl, 10 mM $MgCl_2$, and 100 μg BSA/ml. The optimum storage medium for the enzyme is 10 mM Tris-HCl, pH 7.5, 50 mM KCl, 1 mM DTT, 0.1 mM EDTA, 300 μg BSA/ml and 50 % glycerol (v/v). The enzyme is stored at –20 °C. The optimum temperature for the enzyme assay is 37 °C. The enzyme is fully active up to 5 hours in the assay system. It gets inhibited by *dam* methylation but not by dcm methylation.

xiii. Dde I

The restriction endonuclease, Dde I has been isolated from *Desulfovibrio desulfuricans*. The recognition sequence and the cutting site within the sequence is shown below:

C ↓ TNAG
GANT ↑ C

The enzyme produces 5' protruding ends. The optimum assay buffer for the enzyme activity is 10 mM Tris-HCl, pH 8.0, 150 mM KCl, 10 mM $MgCl_2$, 10 mM 2-ME and 100 μg BSA/ml. The optimum storage medium for the enzyme is 10 mM Tris-HCl, pH 7.5, 50 mM KCl, 1 mM DTT, 0.1 mM EDTA, 300 μg BSA/ml and 50 % glycerol (v/v). The enzyme is stored at –20 °C. The optimum temperature for the enzyme assay is 37 °C. The enzyme is fully active up to one hour and partially active up to 5 hours in the assay system.

xiv. Dra I

The restriction endonuclease, Dra I has been isolated from *Deinococcus radiophilus*. The recognition sequence and the cutting site within the sequence is shown below:

TTT ↓ AAA

AAA ↑ TTT

The enzyme produces blunt ends. The optimum assay buffer for the enzyme activity is 10 mM Tris-HCl, pH 8.0, 10 mM $MgCl_2$, 1 mM 2-ME and 100 μg BSA/ml. The optimum storage medium for the enzyme is 10 mM Tris-HCl, pH 7.5, 50 mM KCl, 1 mM DTT, 0.1 mM EDTA, 300 μg BSA/ml and 50 % glycerol (v/v). The enzyme is stored at –20 °C. The optimum temperature for the enzyme assay is 37 °C. The enzyme is fully active up to one hour and partially active up to 3 hours in the assay system.

xv. EcoR I

The restriction endonuclease, EcoR I has been isolated from *Escherichia coli* RY13. The recognition sequence and the cutting site within the sequence is shown below:

G ↓ AATTC

CTTAA ↑ G

The enzyme produces 5' protruding ends. The optimum assay buffer for the enzyme activity is 10 mM Tris-HCl, pH 7.5, 100 mM NaCl, 10 mM $MgCl_2$, 1 mM 2-ME and 100 μg BSA/ml. The optimum storage medium for the enzyme is 10 mM Tris-HCl, pH 7.5, 50 mM KCl, 400 mM NaCl, 1 mM DTT, 0.1 mM EDTA, 300 μg BSA/ml, 0.15 % Triton-X-100 and 50 % glycerol (v/v). The enzyme is stored at –20 °C. The optimum temperature for the enzyme assay is 37 °C. The enzyme is fully active up to 5 hours in the assay system. Under certain conditions, the enzyme exhibits star activity.

xvi. Hae II

The restriction endonuclease, Hae II has been isolated from *Haemophilus aegyptius*. The recognition sequence and the cutting site within the sequence is shown below:

↓

A *C*

GGCGC T

C ↑ CGCGA

T G

The enzyme produces 3' protruding ends. The optimum assay buffer for the enzyme activity is 10 mM Tris-HCl, pH 8.0, 50 mM NaCl, 10 mM $MgCl_2$, 10 mM 2-ME and 100 µg BSA/ml. The optimum storage medium for the enzyme is 10 mM Tris-HCl, pH 7.5, 50 mM KCl, 1 mM DTT, 0.1 mM EDTA, 300 µg BSA/ml and 50 % glycerol (v/v). The enzyme is stored at –20 °C. The optimum temperature for the enzyme assay is 37 °C. The enzyme is fully active up to one hour and partially active up to 5 hours in the assay system. If methylation of C (GGmCGCT) is there, the sequence is not cut by the enzyme.

xvii. Hae III

The restriction endonuclease, Hae III has been isolated from *Haemophilus aegyptius*. The recognition sequence and the cutting site within the sequence is shown below:

GG ↓ CC

CC ↑ GG

The enzyme produces blunt ends. The optimum assay buffer for the enzyme activity is 10 mM Tris-HCl, pH 8.0, 50 mM NaCl, 10 mM $MgCl_2$, 10 mM 2-ME and 100 µg BSA/ml. The optimum storage medium for the enzyme is 10 mM Tris-HCl, pH 7.5, 50 mM KCl, 1 mM DTT, 0.1 mM EDTA, 300 µg BSA/ml and 50 % glycerol (v/v). The enzyme is stored at –20 °C. The optimum temperature for the enzyme assay is 37 °C. The enzyme is fully active up to two hours and partially active up to 5 hours in the assay system. If methylation of C (GGmCC) is there, the sequence is not cut by the enzyme. However, if methylation of another C is there (GGCmC), the sequence is cut by the enzyme.

xviii. Hinc II

The restriction endonuclease, Hinc II has been isolated from *Haemophilus influenzae* R_c. The recognition sequence and the cutting site within the sequence is shown below:

↓

C A

GTT GAC

CAA CTG

↑

GT

The enzyme produces blunt ends. The optimum assay buffer for the enzyme activity is 10 mM Tris-HCl, pH 7.5, 60 mM NaCl, 10 mM $MgCl_2$, 1 mM 2-ME and 100 μg BSA/ml. The optimum storage medium for the enzyme is 10 mM Tris-HCl, pH 7.5, 50 mM KCl, 1 mM DTT, 0.1 mM EDTA, 300 μg BSA/ml and 50 % glycerol (v/v). The enzyme is stored at –20 °C. The optimum temperature for the enzyme assay is 37 °C. The enzyme is fully active up to one hour and partially active up to 5 hours in the assay system.

xix. Hind III

The restriction endonuclease, Hind III has been isolated from *Haemophilus influenzae* R_d. The recognition sequence and the cutting site within the sequence is shown below:

A ↓ AGCTT

TTCGA ↑ A

The enzyme produces 5' protruding ends. The optimum assay buffer for the enzyme activity is 10 mM Tris-HCl, pH 8.0, 60 mM NaCl, 10 mM $MgCl_2$, 1 mM 2-ME and 100 μg BSA/ml. The optimum storage medium for the enzyme is 10 mM Tris-HCl, pH 7.5, 300 mM KCl, 1 mM DTT, 0.1 mM EDTA, 300 μg BSA/ml and 50 % glycerol (v/v). The enzyme is stored at –20 °C. The optimum temperature for the enzyme assay is 37 °C. The enzyme is fully active up to two hours and partially active up to 5 hours in the assay system. If methylation of A or C (mAAGCTT, AAGmCTT) is there, the site is not cut by the enzyme.

xx. Hinf I

The restriction endonuclease, Hinf I has been isolated from *Haemophilus influenzae* R_f. The recognition sequence and the cutting site within the sequence is shown below:

G ↓ ANTC

CTNA ↑ G

The enzyme produces 5' protruding ends. The optimum assay buffer for the enzyme activity is 10 mM Tris-HCl, pH 7.5, 50 mM NaCl, 10 mM $MgCl_2$, 1 mM 2-ME and 100 μg BSA/ml. The optimum storage medium for the enzyme is 10 mM Tris-HCl, pH 7.5, 50 mM KCl, 1 mM DTT, 0.1 mM EDTA, 300 μg BSA/ml and 50 % glycerol (v/v). The enzyme is stored at –20 °C. The optimum temperature for the enzyme assay is 37 °C. The enzyme is fully active up to one hour and partially active up to 3 hours in the assay system.

xxi. Hpa I

The restriction endonuclease, Hpa I has been isolated from *Haemophilus parainfluenzae*. The recognition sequence and the cutting site within the sequence is shown below:

GTT ↓ AAC

CAA ↑ TTG

The enzyme produces blunt ends. The optimum assay buffer for the enzyme activity is 10 mM Tris-HCl, pH 7.5, 25 mM KCl, 10 mM $MgCl_2$, 1 mM 2-ME and 100 μg BSA/ml. The optimum storage medium for the enzyme is 10 mM Tris-HCl, pH 7.5, 50 mM KCl, 1 mM DTT, 0.1 mM EDTA, 300 μg BSA/ml and 50 % glycerol (v/v). The enzyme is stored at –20 °C. The optimum temperature for the enzyme assay is 37 °C. The enzyme is fully active up to one hour and partially active up to 5 hours in the assay system. Under certain conditions, the enzyme exhibits star activity. If methylation of C (GTTAAmC) is there, the site is cut by the enzyme. However, if methylation of A (GTTAmAC) is there, the site is not cut by the enzyme.

xxii. Hpa II

The restriction endonuclease, Hpa II has been isolated from *Haemophilus parainfluenzae*. The recognition sequence and the cutting site within the sequence is shown below:

C ↓ CGG

GGC ↑ C

The enzyme produces 5' protruding ends. The optimum assay buffer for the enzyme activity is 10 mM Tris-HCl, pH 7.5, 10 mM KCl, 10 mM $MgCl_2$, 1 mM 2-ME and 100 μg BSA/ml. The optimum storage medium for the enzyme is 10 mM Tris-HCl, pH 7.5, 50 mM KCl, 1 mM DTT, 0.1 mM EDTA, 300 μg BSA/ml and 50 % glycerol (v/v). The enzyme is stored at –20 °C. The optimum temperature for the enzyme assay is 37 °C. The enzyme is fully active up to 5 hours in the assay system. If methylation of C (mCCGG, CmCGG) is there, the site is not cut by the enzyme.

xxiii. Kpn I

The restriction endonuclease, Kpn I has been isolated from *Klebsiella pneumoniae*. The recognition sequence and the cutting site within the sequence is shown below:

GGTAC ↓ C

C ↑ CATGG

The enzyme produces 3' protruding ends. The optimum assay buffer for the enzyme activity is 10 mM Tris-HCl, pH 7.5, 10 mM NaCl, 10 mM $MgCl_2$, 10 mM 2-ME and 100 μg BSA/ml. The optimum storage medium for the enzyme is 10 mM Tris-HCl, pH 7.5, 50 mM KCl, 1 mM DTT, 0.1 mM EDTA, 300 μg BSA/ml and 50 % glycerol (v/v). The enzyme is stored at –20 °C. The optimum temperature for the enzyme assay is 37 °C. The enzyme is fully active up to one hour and partially active up to 2 hours in the assay system.

xxiv. Nar I

The restriction endonuclease, Nar I has been isolated from *Nocardia argentinensis*. The recognition sequence and the cutting site within the sequence is shown below:

GG ↓ CGCC

CCGC ↑ GG

The enzyme produces 5' protruding ends. The optimum assay buffer for the enzyme activity is 10 mM Tris-HCl, pH 7.5, 10 mM KCl, 10 mM $MgCl_2$, 5 mM 2-ME and 100 μg BSA/ml. The optimum storage medium for the enzyme is 10 mM Tris-HCl, pH 7.5, 50 mM KCl, 1 mM DTT, 0.1 mM EDTA, 200 mg BSA/ml and 50 % glycerol (v/v). The enzyme is stored at –20 °C. The optimum temperature for the enzyme assay is 37 °C. The enzyme is fully active up to 5 hours in the assay system. It does not cleave DNA when 5' C residue is methylated but will cleave DNA if 3' C residue is methylated.

xxv. Nco I

The restriction endonuclease, Nco I has been isolated from *Nocardia corallina*. The recognition sequence and the cutting site within the sequence is shown below:

C ↓ CATGG

GGTAC ↑ C

The enzyme produces 5' protruding ends. The optimum assay buffer for the enzyme activity is 10 mM Tris-HCl, pH 8.0, 150 mM NaCl, 10 mM $MgCl_2$, 1 mM 2-ME and 100 μg BSA/ml. The optimum storage medium for the enzyme is 10 mM Tris-HCl, pH 7.5, 50 mM KCl, 1 mM DTT, 0.1 mM EDTA, 300 μg BSA/ml and 50 % glycerol (v/v). The enzyme is stored at –20 °C. The optimum temperature for the enzyme assay is 37 °C. The enzyme is fully active up to 3 hours and partially active up to 5 hours in the assay system.

xxvi. Nde I

The restriction endonuclease, Nde I has been isolated from *Neisseria denitrificans*. The recognition sequence and the cutting site within the sequence is shown below:

CA ↓ TATG

GTAT ↑ AC

The enzyme produces 5' protruding ends. The optimum assay buffer for the enzyme activity is 10 mM Tris-HCl, pH 8.0, 150 mM NaCl, 10 mM $MgCl_2$, 6 mM 2-ME and 100 μg BSA/ml. The optimum storage medium for the enzyme is 10 mM Tris-HCl, pH 7.5, 50 mM KCl, 1 mM DTT, 0.1 mM EDTA, 300 μg BSA/ml and 50 % glycerol (v/v). The enzyme is stored at –20 °C. The optimum temperature for the enzyme assay is 37 °C. The enzyme is fully active up to one hour and partially active up to 5 hours in the assay system. It has been found to be sensitive to impurities in DNA.

xxvii. Pst I

The restriction endonuclease, Pst I has been isolated from *Providencia stuartii*. The recognition sequence and the cutting site within the sequence is shown below:

CTGCA ↓ G

G ↑ ACGTC

The enzyme produces 3' protruding ends. The optimum assay buffer for the enzyme activity is 10 mM Tris-HCl, pH 8.0, 50 mM NaCl, 10 mM $MgCl_2$, 1 mM 2-ME and 100 μg BSA/ml. The optimum storage medium for the enzyme is 10 mM Tris-HCl, pH 7.5, 50 mM KCl, 100 mM NaCl, 1 mM DTT, 0.1 mM EDTA, 300 μg BSA/ml, 0.15 % Triton-X-100 and 50 % glycerol (v/v). The enzyme is stored at –20 °C. The optimum temperature for the enzyme assay is 37 °C. The enzyme is fully active up to one hour and partially active up to 2 hours in the assay system. It has been found to be affected by dcm methylase. It may also show star activity if the conditions vary.

xxviii. Pvu I

The restriction endonuclease, Pvu I has been isolated from *Proteus vulgaris*. The recognition sequence and the cutting site within the sequence is shown below:

CGAT ↓ CG

GC ↑ TAGC

The enzyme produces 3' protruding ends. The optimum assay buffer for the enzyme activity is 10 mM Tris-HCl, pH 8.0, 150 mM KCl, 10 mM $MgCl_2$, 10 mM 2-ME and 100 μg BSA/ml. The optimum storage medium for the enzyme is 10 mM Tris-HCl, pH 7.5, 100 mM KCl, 1 mM DTT, 0.1 mM EDTA, 300 μg BSA/ml and 50 % glycerol (v/v). The enzyme is stored at –20 °C. The optimum temperature for the enzyme assay is 37 °C. The enzyme is fully active up to 3 hours and partially active up to 5 hours in the assay system. There may be up to 5 fold decrease in the enzyme activity at pH 7.5 instead of pH 8.0. If methylation of A (CGmATCG) is there, the site is cut by the enzyme. However, if methylation of C (CGATmCG) is there, the site is not cut by the enzyme.

xxix. Pvu II

The restriction endonuclease, Pvu II has been isolated from *Proteus vulgaris*. The recognition sequence and the cutting site within the sequence is shown below:

CAG ↓ CTG

GTC ↑ GAC

The enzyme produces blunt ends. The optimum assay buffer for the enzyme activity is 10 mM Tris-HCl, pH 7.5, 60 mM NaCl, 10 mM $MgCl_2$, 10 mM 2-ME and 100 μg BSA/ml. The optimum storage medium for the enzyme is 10 mM Tris-HCl, pH 7.5, 50 mM KCl, 1 mM DTT, 0.1 mM EDTA, 300 μg BSA/ml and 50 % glycerol (v/v). The enzyme is stored at –20 °C. The optimum temperature for the enzyme assay is 37 °C. The enzyme is fully active up to one hour and partially active up to 4 hours in the assay system. There may be up to 5 fold decrease in the enzyme activity at pH 7.5 instead of pH 8.0. Its activity may be affected by *dcm* methylase.

xxx. Rsa I

The restriction endonuclease, Rsa I has been isolated from *Rhodopseudomonas sphaeroides*. The recognition sequence and the cutting site within the sequence is shown below:

GT ↓ AC

CA ↑ TG

The enzyme produces blunt ends. The optimum assay buffer for the enzyme activity is 10 mM Tris-HCl, pH 8.0, 50 mM KCl, 10 mM $MgCl_2$, 7 mM 2-ME and 100 μg BSA/ml. The optimum storage medium for the enzyme is 10 mM Tris-HCl, pH 7.5, 50 mM KCl, 1 mM DTT, 0.1 mM EDTA, 300 μg BSA/ml and 50 % glycerol (v/v). The enzyme is stored at –20 °C. The optimum temperature for the enzyme assay is 37 °C. The enzyme is fully active up to one hour and partially active up to 2 hours in the assay system.

xxxi. Sac I

The restriction endonuclease, Sac I has been isolated from *Streptomyces achromogenes*. The recognition sequence and the cutting site within the sequence is shown below:

GAGCT ↓ C

C ↑ TCGAG

The enzyme produces 3' protruding ends. The optimum assay buffer for the enzyme activity is 10 mM Tris-HCl, pH 7.5, 20 mM NaCl, 10 mM $MgCl_2$, 1 mM 2-ME and 100 μg BSA/ml. The optimum storage medium for the enzyme is 10 mM Tris-HCl, pH 7.5, 50 mM KCl, 1 mM DTT, 0.1 mM EDTA, 300 μg BSA/ml and 50 % glycerol (v/v). The enzyme is stored at –20 °C. The optimum temperature for the enzyme assay is 37 °C. The enzyme is fully active at least up to 2 hours in the assay system.

xxxii. Sac II

The restriction endonuclease, Sac II has been isolated from *Streptomyces achromogenes*. The recognition sequence and the cutting site within the sequence is shown below:

↓
CCGC GG
GG CGCC
↑

The enzyme produces 3' protruding ends. The optimum assay buffer for the enzyme activity is 10 mM Tris-HCl, pH 7.5, 10 mM KCl, 10 mM $MgCl_2$, 10 mM 2-ME and 100 μg BSA/ml. The optimum storage medium for the enzyme is 10 mM Tris-HCl, pH 7.5, 50 mM KCl, 1 mM DTT, 0.1 mM EDTA, 300 μg BSA/ml and 50 % glycerol (v/v). The enzyme is stored at –20 °C. The optimum temperature for the enzyme assay is 37 °C. The enzyme is fully active at least up to 2 hours in the assay system.

xxxiii. Sal I

The restriction endonuclease, Sal I has been isolated from *Streptomyces albus* G. The recognition sequence and the cutting site within the sequence is shown below:

↓
G TCGAC
CAGCT G
↑

The enzyme produces 5' protruding ends. The optimum assay buffer for the enzyme activity is 10 mM Tris-HCl, pH 8.0, 150 mM NaCl, 10 mM $MgCl_2$, 1 mM 2-ME and 100 μg BSA/ml. The optimum storage medium for the enzyme is 10 mM Tris-HCl, pH 7.5, 100 mM KCl, 1 mM DTT, 0.1 mM EDTA, 300 μg BSA/ml and 50 % glycerol (v/v). The enzyme is stored at –20 °C. The optimum temperature for the enzyme assay is 37 °C. The enzyme is fully active up to 2 hours and partially active up to 5 hours in the assay system. The enzyme may exhibit star activity if the conditions of the enzyme assay vary. If methylation of A or C (GTCGmAC, GTmCGAC) is there, the site is not cut by the enzyme.

xxxiv. Sma I

The restriction endonuclease, Sma I has been isolated from *Serratia marcescens*. The recognition sequence and the cutting site within the sequence is shown below:

CCC ↓ GGG

GGG ↑ CCC

The enzyme produces blunt ends. The optimum assay buffer for the enzyme activity is 10 mM Tris-HCl, pH 8.0, 20 mM KCl, 10 mM $MgCl_2$, 10 mM 2-ME and 100 μg BSA/ml. The optimum storage medium for the enzyme is 10 mM Tris-HCl, pH 7.5, 150 mM KCl, 1 mM DTT, 0.1 mM EDTA, 300 μg BSA/ml, 0.15 % Triton-X-100 and 50 % glycerol (v/v). The enzyme is stored at –20 °C. The optimum temperature for the enzyme assay is 30 °C. The enzyme is fully active up to one hour and partially active up to 4 hours in the assay system. If methylation of C (CCmCGGG) is there, the site is not cut by the enzyme. However, if methylation of C (CmCmCGGG) is there, the site is cut by the enzyme.

xxxv. Taq I

The restriction endonuclease, Taq I has been isolated from *Thermus aquaticus*. The recognition sequence and the cutting site within the sequence is shown below:

T ↓ CGA

AGC ↑ T

The enzyme produces 5' protruding ends. The optimum assay buffer for the enzyme activity is 10 mM Tris-HCl, pH 8.4, 100 mM NaCl, 10 mM $MgCl_2$, 10 mM 2-ME and 100 μg BSA/ml. The optimum storage medium for the enzyme is 10 mM Tris-HCl, pH 7.5, 300 mM KCl, 1 mM DTT, 0.1 mM EDTA, 300 μg BSA/ml and 50 % glycerol (v/v). The enzyme is stored at –20 °C. The optimum temperature for the enzyme assay is 65 °C. The assay mixture is generally incubated under the paraffin oil to avoid evaporation. The enzyme is fully active up to 3 hours and partially active up to 5 hours in the assay system. If methylation of C (TmCGA) is there, the site is cut by the enzyme. However, if methylation of A (TCGmA) is there, the site is not cut by the enzyme.

xxxvi. Xba I

The restriction endonuclease, Xba I has been isolated from *Xanthomonas badrii*. The recognition sequence and the cutting site within the sequence is shown below:

T ↓ CTAGA

AGATC ↑ T

The enzyme produces 5' protruding ends. The optimum assay buffer for the enzyme activity is 10 mM Tris-HCl, pH 8.0, 100 mM NaCl, 10 mM $MgCl_2$, 1 mM 2-ME and 100 μg BSA/ml. The optimum storage medium for the enzyme is 10 mM Tris-HCl, pH 7.5, 200 mM

KCl, 1 mM DTT, 0.1 mM EDTA, 300 μg BSA/ml and 50 % glycerol (v/v). The enzyme is stored at –20 °C. The optimum temperature for the enzyme assay is 37 °C. The enzyme is fully active up to 2 hours and partially active up to 5 hours in the assay system. If methylation of C (TmCTAGA) is there, the site is not cut by the enzyme. The enzyme may exhibit star activity if conditions of the enzyme assay vary.

xxxvii. Xho I

The restriction endonuclease, Xho I has been isolated from *Xanthomonas holcicola.* The recognition sequence and the cutting site within the sequence is shown below:

C ↓ TCGAG

GAGCT ↑ C

The enzyme produces 5' protruding ends. The optimum assay buffer for the enzyme activity is 10 mM Tris-HCl, pH 7.5, 150 mM NaCl, 10 mM $MgCl_2$, 10 mM 2-ME and 100 μg BSA/ml. The optimum storage medium for the enzyme is 10 mM Tris-HCl, pH 7.5, 50 mM KCl, 1 mM DTT, 0.1 mM EDTA, 300 μg BSA/ml and 50 % glycerol (v/v). The enzyme is stored at –20 °C. The optimum temperature for the enzyme assay is 37 °C. The enzyme is fully active up to 5 hours in the assay system. If methylation of C or A (CTmCGAG, CTCGmAG) is there, the site is not cut by the enzyme.

3.3. DNA Methylases

There are two common enzymes called as *dam* methylase and *dcm* methylase present in many *E.coli* strains. These enzymes methylate adenine residue and cytosine residue, respectively in specific sequences.

i. *dam* Methylase

The enzyme transfers methyl group from S-adenosyl methionine on to N^6 of adenine residue when present in a specific sequence, GATC. This sequence is present in the recognition sequences of many restriction endonuclease and in most of the cases, after methylation, the site is not cut by the restriction endonuclease.

ii. *dcm* Methylase

The enzyme transfers methyl group from S-adenosyl methionine on to cytosine residue forming 5-methyl cytosine when cytosine present in a specific sequence, CCAGG or CCTGG. The sequence is present in the recognition sequences of many restriction endonucleases and after methylation, generally, the site is not cut by the restriction endonuclease.

Besides, there is one specific methylase called as **EcoRI methylase**, present in *E.coli* strain RY13 which methylates recognition sequence of EcoRI restriction endonuclease forming GAmATTC which is not cut by EcoRI. Thus EcoRI methylase protects recognition sequence of EcoRI from digestion by EcoRI restriction endonuclease. The enzyme is useful in protecting internal EcoRI sites of the cloning fragments prior to cloning into an EcoRI site. Use of this methylase is taken while cloning cDNA fragments into λgt 10 or 11 vectors. First, all the EcoRI sites are blocked by methylation and afterwards, EcoRI linkers are ligated.

There is an another methylase called as **Hpa II methylase**, isolated from *Haemophilus parainfluenzae* which methylates recognition sequence of Hpa II restriction endonuclease forming CmCGG which is not cut by Hpa II restriction endonuclease. Thus, the methylase protects the recognition site from digestion by Hpa II. Hpa II sites are considered to be involved in the regulation of eukaryotic gene expression, as the site contains the dinucleotide CpG. The enzyme has been used *in vitro* to study eukaryotic methylation patterns. It is used in TAB linker mutagenesis. In a partially Hpa II digested DNA, TAB linker may be inserted at a number of different sites within the gene. Hpa II methylase protects undigested Hpa II sites where a TAB linker has not been inserted.

References

[1] Roberts, R.J. (1983) Restriction and modification enzymes and their recognition sequences. *Nucleic Acids Res.* **11**: r135-167.

[2] Roberts, R.J., Myers, P.A., Morrison, A. and Murray, K. (1976) A specific endonuclease from Haemophilus haemolyticus. *J Mol Biol.* **103**:199-208.

[3] Murray, K., Hughes, S.G., Brown, J.S. and Bruce, S.A. (1976) Isolation and characterization of two sequence-specific endonucleases from Anabaena variabilis. *Biochem J.* **159**:317-322.

[4] Wilson, G.A. and Young, F.E. (1975) Isolation of a sequence-specific endonuclease (BamI) from *Bacillus amyloliquefaciens* H. *J Mol Biol.* **97**:123-125.

[5] Bingham, A.H., Atkinson, T., Sciaky, D. and Roberts, R.J. (1978) A specific endonuclease from Bacillus caldolyticus. *Nucleic Acids Res.* **5**: 3457-3467.

[6] Lautenberger, J.A., Edgell, M.H. and Hutchison, C.A. (1980) The nucleotide sequence recognized by the BstEII restriction endonuclease. *Gene* **12**:171-174.

[7] Mayer, H., Grosschedl, R., Schutte, H. and Hobom, G. (1981) ClaI. a new restriction endonuclease from Caryophanon latum L. *Nucleic Acids Res.* **9**: 4833-4845.

[8] Makula, R.A. and Meagher,R.B. (1980) A new restriction endonuclease from the anaerobic bacterium, Desulfovibrio desulfuricans, Norway. *Nucleic Acids Res.* **8**:3125-3131.

[9] Landy, A., Ruedisueli, E., Robinson, L., Foeller, C. and Ross W. (1974) Digestion of deoxyribonucleic acids from bacteriophage T7, lambda, and phi 80h with site-specific nucleases from Hemophilus influenzae strain Rc and strain Rd. *Biochemistry* **13**:2134-2142.

[10] Old, R., Murray, K. and Boizes, G. (1975) Recognition sequence of restriction endonuclease III from Hemophilus influenzae. *J Mol Biol.* **92**:331-339.

[11] Sharp, P.A., Sugden, B. and Sambrook, J. (1973) Detection of two restriction endonuclease activities in Haemophilus parainfluenzae using analytical agarose--ethidium bromide electrophoresis. *Biochemistry* **12**: 3055-3063.

[12] Lynn, S.P., Cohen, L.K., Kaplan, S. and Gardner, J.F. (1980) RsaI: a new sequence-specific endonuclease activity from Rhodopseudomonas sphaeroides. *J Bacteriol.* **142**: 380-383.

[13] Arrand, J.R., Myers, P.A. and Roberts, R.J. (1978) A new restriction endonuclease from Streptomyces albus G. *J Mol Biol.* **118**: 127-135.

[14] Endow, S.A. and Roberts, R.J. (1977) Two restriction-like enzymes from Xanthomonas malvacearum. *J Mol Biol.* **112**: 521-529.

[15] Zain, B.S. and Roberts, R.J. (1977) A new specific endonuclease from Xanthomonas badrii. *J Mol Biol.* **115**: 249-255.

[16] Gingeras, T.R., Myers, P.A., Olson, J.A., Hanberg, F.A. and Roberts, R.J. (1978) A new specific endonuclease present in Xanthomonas holcicola, Xanthomonas papavericola and Brevibacterium luteum. *J Mol Biol.* **118**: 113-122.

[17] Sato, S., Hutchinson, C.A. and Harris, J.I. (1977) A thermostable sequence-specific endonuclease from Thermus aquaticus. *Proc Natl Acad Sci U S A.* **74**: 542-546.

[18] Rubin, R.A. and Modrich, P. (1977) EcoRI methylase. Physical and catalytic properties of the homogeneous enzyme. *J Biol Chem.* **252**: 7265-7272.

Chapter IV

Enzymes Used in Genetic Engineering

4.1. Introduction

There are many enzymes used in molecular cloning and genetic engineering. DNA polymerase I and Klenow fragment are used for filling the gap reactions and replication. Nowadays, heat stable DNA polymerase is used in PCR reaction. DNA polymerase is also used during dideoxy DNA sequencing for extending the growing complementary chain of DNA. Klenow fragment, the larger fragment of DNA polymerase I, and devoid of 5',3' exonuclease activity, is used for replication of the complimentary DNA chain during oligonucleotide site directed mutagenesis. Reverse transcriptase, capable of synthesizing DNA from RNA, has become important for synthesizing complementary DNA (cDNA) from mRNA for making cDNA library. Polynucleotide kinase transfers γ-phosphoryl group from ATP on to 5' –OH end of DNA and is also capable to exchange γ-phosphoryl group of ATP with the phosphoryl group present at the 5' end of the DNA. The enzyme is used for making radioactive probes as well as for labeling DNA fragments for further sequencing by the chemical modification method of Maxam and Gilbert. Terminal deoxynucleotidyl transferase is capable of transferring deoxyribonucleotides onto the 3' ends of DNA strands in a template independent reaction using deoxynucleoside triphosphates. It is used for labeling 3' ends of DNA and labeled DNA is further used for sequencing purposes by the method based on chemical modification method of Maxam and Gilbert. Alkaline phosphatase removes phosphate group from phosphoryl esters. In molecular biology, it is used to dephosphorylate DNA at the 5' ends. S1 Nuclease degrades single stranded nucleic acids. Double stranded DNA, RNA and DNA-RNA hybrids are resistant to its digestion. In molecular cloning, S1 nuclease is used to remove single stranded tails from double stranded DNA to produce blunt ends. It is also used to open hairpin loop structures formed during synthesis of double stranded cDNA from mRNA. DNA ligase is used for end to end joining of the DNA, for cloning DNA fragments in the vectors. In this chapter, emphasis is being given on the practical protocols in addition to brief description of the structure and properties of the enzymes.

4. 2. DNA Polymerase I

All the DNA polymerases have same type of DNA synthetic activity. DNA polymerase extends a DNA chain by adding nucleotides one at a time to the 3'-OH end. DNA polymerase improves the specificity of the complementary strand synthesis by specifically recognizing nucleotides through two mechanisms namely pre-synthetic error control and proof reading control (scrutinization of the base pairs after the new nucleotide addition to the growing chain. All the bacterial DNA polymerases have been found to have 3'-5' exonuclease activity that proceeds in the reverse direction from the DNA synthesis.

E.coli DNA polymerase I is commercially available and is used in molecular cloning and genetic engineering. I*n vivo*, this enzyme is used for DNA repairs. It is coded by the locus *pol*A gene and is a single polypeptide of about 105,000 molecular weight. It exhibits 5',3' exonuclease as well as 3',5' exonuclease activities in addition to DNA dependent DNA polymerase activity. On digestion by trypsin, the enzyme is cleaved in to two fragments: larger fragment of molecular weight nearly 68,000 having DNA dependent DNA polymerase and 3',5' exonuclease activities called Klenow Fragment, and smaller fragment of molecular weight nearly 36,000 having 5',3' exonuclease activity. The 5',3' exonuclease activity excises small groups of nucleotides (up to 10 nucleotides at a time) and this activity is coordinated with the proofreading activity. This enzyme has been used for labeling the DNA and for nick translation *in vitro* by incorporating radioactive labeled nucleotides. The displaced strand is degraded by the 5',3' exonuclease activity of the enzyme. The enzyme requires intact sulfhydryl group for enzyme activity. The enzyme shows optimum pH in the range of pH 7.4. It requires the presence of the template absolutely and also of all the four deoxyribonucleoside triphosphates. The enzyme also requires Mg^{+2} for enzyme activity. The enzyme is nowadays mainly used for nick translation. Although the practical protocol may differ and should be used as per recommendation of the manufacturer, a general assay may be as follows:

- 130 mM Phosphate buffer, pH 7.4
- 10 mM $MgCl_2$
- 1 mM DTT
- 30 μM of each of the four dNTPs
- 1-2 Units of the enzyme
- Incubate at 37°C for 30 min.

Although the enzyme DNA polymerase I was being used for dideoxy DNA sequencing but nowadays, Klenow fragment is preferred.

4.3. Klenow Fragment

It is the large fragment of DNA dependent DNA polymerase I of *E.coli*. It has DNA dependent DNA polymerase and 3',5' exonuclease activities but lacks 5',3' exonuclease activity. The two activities reside in different regions of the fragment. The C-terminal two third of the fragment contains the DNA polymerase activity whereas one third N-terminal contains the 3',5' exonuclease activity that is also called as proofreading activity due to the reason it keeps check on the incorporation of mismatches in the growing DNA chain. It removes those nucleotides that mismatch with the template lowering the error rate to about 10^{-6}. Without proofreading activity, the error rate is found to be nearly 10^{-4} per nucleotide replicated. Klenow fragment has been crystallized. With the help of X-ray diffraction studies, at a resolution of 2.75 A°, a cleft is found for accomodating the primer-template and the site for the 3',5' exonuclease activity is about 30 A° away. Using crystallographic studies, it has been interpreted that single stranded DNA binds near the growing end site. Genetically engineered Klenow fragment is available commercially. Klenow fragment has been used in the DNA sequencing when dideoxy method of Sanger is used. It has also been used in *in vitro* oligonucleotide site directed mutagenesis after annealing the primer having desired mismatch. Nowadays, Klenow fragment is also used for the synthesis of the second strand of cDNA synthesized from the mRNA with the help of reverse transcriptase for further making the cDNA library. It has also been used for the fill in the gap reaction. If a double stranded DNA contains 3' recessive ends and one wishes to generate blunt ends, use of the fill in the gap reaction is taken. Besides, like DNA polymerase I, Klenow fragment is also used for preparing single stranded radioactive labeled probes. One unit of the enzyme activity for Klenow fragment is defined as the amount of the enzyme needed for incorporation of 10 nanomoles of total deoxynucleotides into acid insoluble product in 30 minutes at 37°C with poly(dAdT).ploy(dAdT) as template. The optimum pH has been reported in the range of pH 7.5 to 8.0. A general assay is same as given above for the DNA polymerase I. However, the assay system should be prepared as per recommendation of the manufacturer.

4.4. Reverse Transcriptase

Reverse transcriptase catalyzes the synthesis of DNA using RNA as a template. The enzyme has been found in RNA viruses (retroviruses). After infection, reverse transcriptase of the virus synthesizes cDNA from viral RNA. The cDNA after making second strand with the help of DNA polymerase, becomes integrated into the genome of the host and is inherited like other host genes. This cDNA of the virus, sometimes called as provirus, is transcribed to produce viral RNA by the host RNA polymerase. It is believed that many interspersed repetitive sequences present in the eukaryotic genomes might have integrated by a similar mechanism. Unlike RNA polymerase, reverse transcriptase can not initiate polynucleotide synthesis without a primer. It uses tRNA as a primer that binds through complementary base pairing to the 3' end of the viral RNA. After the synthesis of cDNA, the viral RNA in the DNA-RNA hybrid is hydrolyzed by RNase H, an enzyme activity present in reverse transcriptase itself. Afterwards, the same enzyme (reverse transcriptase) or DNA polymerase may synthesize complimentary DNA strand to give double stranded DNA which becomes integrated into the host genome. The reverse transcriptase of Human immunodeficiency virus (HIV) has been studied and found to be a dimeric protein having two identical size subunits of 560 amino acids. The RNase H activity has been found located with in the carboxyl terminal end and may be removed by restricted proteolysis. Reverse transcriptase is considered to be an important enzyme in genetic engineering. It is used for the synthesis of cDNA from mRNA. Afterwards, using cDNA as template and with the help of DNA polymerase, another complementary strand is synthesized resulting in double stranded cDNA. The double stranded cDNA is used for constructing cDNA library in a suitable vector. The commercially available reverse transcriptase is Avian Myeloblastosis Virus (AMV) reverse transcriptase. It also shows 5',3' primer dependent polymerase activity using either RNA or DNA as a template and 3',5' RNase H activity. The enzyme is consisted of two different types of subunits, α subunit having the molecular weight 62,000 and β subunit having the molecular weight 94,000. There are evidences indicating that α subunit is generated as a result of proteolysis of β subunit. Generally, one unit of reverse transcriptase is defined as the amount of the enzyme needed for incorporation of one nanomole of dTMP into acid insoluble product in 10 minutes at 37°C using poly(rA).poly$(dT)_{12-18}$ as template/primer. A general protocol of reverse transcriptase for synthesizing cDNA may be as follows. For enzyme assay of reverse transcriptase, advantage of Poly A tail in eukaryotic mRNA is taken.

- 10 x buffer (0.5 M Tris- HCl, pH 8.5 containing 0.4 M KCl, 0.1 M $MgCl_2$ and 0.005 M DTT)

Stock solutions of all the four dNTPs (dATP, dTTP, dGTP, dCTP) should be diluted to get final concentration 10 mM.

- 2 μg polyA RNA in 34 μl of H_2O
- 5 μl 10x buffer
- 5 μl 10 mM d(ACGT)TP
- 5 μl oligo $(dT)_{12-18}$ (100 μg/ml)
- 1 0-20 units of reverse transcriptase

Incubate at 37°C or 42 °C for one hour

Denature mRNA-cDNA duplex by heating the contents in a boiling water bath for 2 minutes, cool on ice. Sometimes, preparation of reverse transcriptase has non-specific RNase activity. It should be checked before use for cDNA synthesis otherwise the presence of RNase activity may hydrolyze mRNA. Reverse transcriptase from Molony Murine Leukemia virus is also available commercially. It contains only one type of subunit of the molecular weight 71,000. Like AMV enzyme, it also requires Mg^{+2} for activity.

4.5. Polynucleotide Kinase

Polynucleotide kinase transfers γ-phosphoryl group from ATP on to the 5' ends of DNA molecule. The enzyme is used for labeling 5'-end of the DNA (for use as probe). Besides, 5'-end labeled DNA is also used for DNA sequencing in chemical modification method of Maxam and Gilbert. The dephosphorylated DNA (having –OH group at the 5' end) is the preferred substrate for polynucleotide kinase. Different protocols have been described for

DNAs having 5' protruding ends and blunt or recessive ends. Polynucleotide kinase is also capable of exchanging 5'- phosphoryl group present in DNA with that of γ-phosphoryl group of ATP. It also transfers γ-phosphoryl group from ATP on to the 5' –hydroxyl end of single stranded DNA. If γ-phosphoryl group of ATP is labeled with radio active ^{32}P or ^{33}P, 5' end of DNA would become labeled after polynucleotide kinase reaction. Commercially, genetically engineered T_4 polynucleotide kinase is available. One unit of the polynucleotide kinase enzyme activity is defined as the amount of the enzyme that catalyzes the transfer of one nanomole of phosphate from ATP on to the polynucleotide in 30 minutes at 37°C. The following protocol may be used for labeling 5' ends of DNA:

For Blunt or Recessive Ends

Add the following components in an eppendorf tube:

- 5 μl dephosphorylated DNA 1-50 pmole 5' ends (dephosphorylation of DNA is done using alkaline phosphatase)
- 35 μl 20 mM Tris-HCl, pH 9.5 containing 1 mM spermidine and 0.1 mM EDTA

Mix and incubate at 90°C for 2-3 minutes, afterwards quickly chill on ice and immediately the following components should be added:

- 5μl 0.5 M Tris-HCl, pH 9.5 containing 0.1 M $MgCl_2$, 0.05 M DTT and 50%
- (v/v) glycerol
- 5 μl γ-^{32}P ATP having nearly 1000 Ci/mmole
- 1 μl T_4 polynucleotide kinase, 20 units

Mix and incubate at 37°C for 15 minutes.

For 5' Protruding Ends

- 40 μl dephosphorylated DNA 1-50 pmole 5' ends and water
- 5μl 0.5 M Tris-HCl, pH 7.6 containing 0.1 M $MgCl_2$, 0.05 M DTT, 1 mM spermidine and 1 mM EDTA
- 5 μl γ-^{32}P ATP having nearly 1000 Ci/mmole
- 1 μl T_4 polynucleotide kinase, 20 units

Mix and incubate at 37°C for 30 minutes

For Exchange Reaction with 5' Protruding Ends

- 30 μl of DNA having 5' terminal phosphoryl protruding ends 1-50 pmole and water
- 5 μl 0.5 M Imidazole- HCl, pH 6.6 containing 0.1 M $MgCl_2$, 0.05 M DTT, 1 mM spermidine and 1 mM EDTA

- 3μl 1 mM ADP
- 5 μl γ-^{32}P ATP having nearly 1000 Ci/mmole
- 1 μl T_4 polynucleotide kinase, 20 units

Mix and incubate at 37°C for 30 minutes.
In all the above three cases, further proceed as follows:

- 200 μl 2.5 M Ammonium acetate (chilled)
- 1 μl tRNA (1 mg/ml)
- 750 μl chilled 95 % ethanol

Close the lid and mix by inverting the tube. Incubate in dry ice ethanol bath for 15-20 minutes. Afterwards, centrifuge at 11,000 x *g* for 10 minutes in a microfuge in cold condition. Remove the supernatant with the help of a pasteur pipet carefully. The tube contains much higher amount of radioactivity. All the possible care must be taken. To the pellet, the following is added:

- μl 0.3 M sodium acetate

Close the lid and dissolve the pellet (DNA) by mild vortexing taking care for radioactivity and add:

- 750 μl chilled 95 % ethanol

Close the lid and mix by inverting the tube. Incubate in dry ice ethanol bath for 15-20 minutes. Following incubation, centrifuge at 11,000 x *g* for 10 minutes in a microfuge in cold condition. Remove the supernatant with the help of a pasteur pipet carefully. To the pellet, add:

- 1 ml chilled 95 % ethanol for washing the pellet

Close the lid and mix by inverting the tube. Centrifuge at 11,000 x *g* for 10 minutes in a microfuge in cold conditions. Remove the supernatant with the help of a pasteur pipet carefully. Remove the traces of the liquid with the help of a pipetman. Dissolve the pellet DNA in 50 μl of water. It contains 5' end labeled DNA. One μl of the solution may be used for checking the amount of the radioactivity in the scintillation counter which is indicative of the labeling.

Glycerol stabilizes the enzyme activity. Spermidine stimulates incorporation of the phosphoryl group at the 5' end and also inhibits nuclease if present in some polynucleotide kinase preparations. If active nuclease is there, DNA will be chewed by it. The concentration of ATP in the reaction mixture should be at least 1 μM. DNA molecules having recessive 5' ends may require higher concentration of ATP especially if denaturation is not effective. Ammonium acetate promotes DNA renaturation and also prevents the precipitation of protein while enhancing DNA precipitation by ethanol. tRNA acts as carrier DNA and gets precipitated out alongwith DNA and gets adhered to the wall of the tube after centrifugation.

Nowadays, many manufacturers provide ready-made reaction buffer. If so, it should be preferred to use ready-made reaction buffer as per directions of the manufacturer.

4.6. Terminal Deoxynucleotidyl Transferase

Terminal deoxynucleotidyl transferase transfers deoxyribonucleotides on to the 3' ends of DNA strands. The reaction is template independent using deoxyribonucleoside triphosphate as precurssor. If deoxyribonucleoside triphosphate has radioactive labeled element (say ^{32}P or ^{33}P) in its α-phosphoryl position, 3' ends of DNA will get labeled. If the reaction is carried out for a longer time and the product is hydrolyzed in the alkaline condition, the DNA 3' ends retain two labeled phosphates, one on either side of the initial deoxyribonucleotide added. The enzyme is a type of DNA polymerase found only in prelymphocytes and in early stages of lymphoid differentiation. In genetic engineering, the enzyme is used to label 3' ends of DNA and labeled DNA is further used for DNA sequencing in the chemical modification reactions of Maxam and Gilbert. The enzyme is commercially available. The following protocol given by Roychouchury and Wu may be used for labeling 3' ends (protruding or blunt ends). Since 3'-end of DNA has hydroxyl group and not phosphoryl group, there is no need of dephosphorylation before reaction of terminal deoxynucleotide transferase. The following components may be added in an eppendorf tube:

- 15 µl DNA, 1-50 pmole 3' ends

Incubate at 90°C for 2 minutes, quickly chill on ice and centrifuge for few seconds. Afterwards, add:

- µl 1 M potassium cacodylate, pH 7.6 containing 0.01 M $CoCl_2$ and 2 mM DTT
- 5 µl α-^{32}P dNTP, nearly 3000 Ci/mmole
- 1 µl Terminal deoxynucleotidyl transferase, 10-20 units

Mix and incubate at 37°C for 30 minutes, add:

- 1 µl Terminal deoxynucleotidyl transferase, 10-20 units

Mix and incubate at 37°C for 30 minutes, add:

- µl 100 mM dNTP

Mix and incubate at 37 °C for 10 minutes, and chill on ice, add:

- 200 µl 2.5 M Ammonium acetate (chilled)

- 1 μl tRNA (1 mg/ml)
- 750 μl chilled 95 % ethanol

Close the lid and mix by inverting the tube. Incubate in dry ice ethanol bath for 15-20 minutes. Afterwards, centrifuge at 11,000 x *g* for 10 minutes in a microfuge in cold condition. Remove the supernatant with the help of a pasteur pipet carefully. The tube contains much higher amount of radioactivity. All the possible care must be taken. To the pellet, the following is added:

- 250 μl 0.3 M sodium acetate

Close the lid and dissolve the pellet (DNA) by mild vortexing taking care for radioactivity and add:

- 750 μl chilled 95 % ethanol

Close the lid and mix by inverting the tube. Incubate in dry ice ethanol bath for 15-20 minutes. Afterwards, centrifuge at 11,000 x *g* for 10 minutes in a microfuge in cold condition. Remove the supernatant with the help of a pasteur pipet carefully. To the pellet, add:

- 1 ml chilled 95 % ethanol for washing the pellet

Close the lid and mix by inverting the tube. Centrifuge at 11,000 x *g* for 10 minutes in a microfuge in cold conditions. Remove the supernatant with the help of a pasteur pipet carefully. Remove the traces of the liquid with the help of a pipetman. Dissolve the DNA pellet in 50 μl of water. It contains 3' end labeled DNA. One μl of the solution may be used for checking the amount of the radioactivity in the scintillation counter which is indicative of the labeling.

The enzyme may also be used for adding complementary homopolymeric tails to vectors and cDNA.

4.7. Alkaline Phosphatase

Alkaline phosphatase hydrolyzes phosphate ester bond releasing inorganic phosphate. Alkaline phosphatases from various sources have been found to exhibit optimum pH in the range of 8 to 10. The name, alkaline phosphatase has been coined due to alkaline pH of the reaction and hydrolysis of the phosphate ester bond by the enzyme. It is a group specific enzyme and therefore may use any phosphoryl ester as its substrate. Commercially different molecular biology grade alkaline phosphatases are available, viz. bacterial alkaline phosphatase (BAP) and calf intestine alkaline phosphatase (CIAP). One unit of the enzyme is defined as the amount of the enzyme needed to hydrolyze one micromole of p-nitro phenyl phosphate (a synthetic substrate) per minute at 37°C in case of CIAP or at 65°C in case of

BAP. The optimum temperatures of CIAP and BAP have been found to be 37°C and 65°C, respectively. Alkaline phosphatase is used to dephosphorylate vector linear DNA in order to avoid self ligation as a result of DNA ligase reaction and also to dephosphorylate linear DNA molecules for transfer of γ-labeled phosphoryl group of ATP in the presence of polynucleotide kinase. A general protocol for alkaline phosphatase is as follows. However, the protocol recommended by the supplier should be followed by the user:

- 50 µl DNA (1-2 µg) in 0.05 M Tris-HCl, pH 8.0 containing 0.001 M EDTA
- 6 µl alkaline phosphatase (average 50 units per µg DNA)

Incubate at 65°C in case of BAP and at 37°C in case of CIAP for 30 minutes.

Some researchers prefer to add few more units of alkaline phosphatase and incubate for 15- 30 minutes additionally.

Add 50 µl more of the buffer and equal volume of Phenol-Chloroform-Isoamylalcohol (ratio 100: 96: 04), vortex gently to mix and keep for few minutes to separate phenol and aqueous layers (Phenol has to be processed before use using a method called as Sewag procedure for phenol preparation as described above in this chapter). The upper aqueous layer is carefully removed with the help of a pipetman and transferred in an another eppendorf tube. Again add equal volume of Phenol-Chloroform-Isoamylalcohol and repeat the process. Aqueous layer containing DNA is extracted with water saturated diethyl ether to extract traces of phenol if any in the preparation. The step is repeated at least twice. Chloroform may also be used in place of diethyl ether. Ether traces are removed by incubating the tubes at 37°C in a water bath for few minutes. Either NaCl to a final concentration of 0.2 M or sodium acetate to a final concentration of 0.25 M is added followed by two to three volumes of chilled 95 % ethanol and allowed to cool in dry ice-ethanol bath for 15-20 minutes. Centrifuge at 11,000 x *g* for 10 minutes in the cold condition. After removing the supernatant, the DNA pellet is washed with 95 % chilled ethanol and dried under vaccum using centrifugal lyophilizer. Dried pellet may be dissolved in 20-30 µl of water.

4.8. S1 Nuclease

S1 Nuclease degrades single stranded nucleic acids to yield 5'-phosphoryl mono-or oligonucleotides. S1 nuclease is unable to chew double stranded DNA, RNA and DNA-RNA hybrids. It has been used to analyze DNA-RNA hybrid structures to map transcripts. It is also used to remove single stranded tails from DNA fragments to produce blunt ends which may be used for ligation with specific linkers. One unit of S1 nuclease is defined as the amount of the enzyme needed for the liberation of one microgram of acid soluble deoxyribonucleotides at 37°C in one minute after incubation with single stranded DNA. The following general protocol may be used:

- 5x buffer: 150 mM sodium acetate buffer pH 4.4-4.6, containing 50 mM NaCl and 5 mM $ZnCl_2$

- 100 μl single stranded DNA (about 500 μg) plus water
- 30 μl 5x buffer
- 20 μl S1 nuclease (about 100 units)

Incubate at 37°C for 30 minutes.

Reaction may be stopped by the addition of phenol-chloroform-isoamylalcohol.

Further procedure of alcoholic precipitation of DNA may be carried out in the same way as described above.

4.9. DNA Ligase

DNA ligase is used for end to end joining of linear DNA molecules. The enzyme from *E.coli* has been found to have a single polypeptide of the molecular weight nearly 77,000 daltons. About 200-400 copies of the enzyme are present in each *E.coli* cell. DNA ligase from *E.coli* is NAD^+ dependent. During the reaction, an adenylyl group is transferred from NAD^+ on to an ε- amino group of lysine present in the enzyme protein. Adenylyl group functions in an energy transfer manner and is linked by a phosphoamide bond with the ε-amino group of lysine.

$E + NAD^+$ ------------------ E-AMP + NMN^+

In eukaryotic cells and some phages, ATP dependent DNA ligase is found where ATP transfers its adenylyl group instead of NAD^+

E + ATP ------------------- E-AMP + PPi

Upon binding of adenylyl group on to the ε- amino group of lysine, it is transferred to the 5' phosphate end of one DNA strand to form a new pyrophosphate bond. Thereafter, 3'-hydroxyl terminus group of the other DNA strand displaces the adenylyl group from the first DNA strand to form a new phosphodiester linkage resulting in joining of 5' terminal and 3' terminal ends of two strands of DNA. DNA ligase has also been used for repairs of single stranded nicks in duplex DNA. Owing to its end to end joining property, it is used for converting linear DNA into circular DNA. It is also used during replication with DNA polymerase (for joining the Okazaki fragments). It has also been used to join segments of DNA in DNA recombination which occurs during genetic transformation, transduction and lysogenization in bacteria and during meiosis in eukaryotes. In genetic engineering, DNA ligase is used for cloning the DNA fragment in the vector. For cloning, any blunt end containing DNA may be ligated with any other blunt ends containing DNA. However, cohesive ends containing DNA may be ligated only with the same type cohesive ends containing DNA. DNA ligase is also used for joining linkers to the blunt ends having DNA. Commercially, ATP dependent T4 DNA ligase is available. Polyethylene glycol has been shown to stimulate blunt end ligation with T4 DNA ligase. It has been shown that ligation is good if the reaction mixture is incubated at 12°C for 16-24 hours. However, some researchers have shown comparable results with incubation at 23°C for 5 hours.

Nowadays, some suppliers supply 2X quick ligase buffer and recommend incubation of the reaction mixture with DNA ligase only for 15 minutes at the room temperature (25°C).

The reaction mixture contains plasmid DNA (linear dephosphorylated) and DNA to be ligated (both plasmid and DNA to be ligated should have same type of cohesive ends). However, any blunt end can be ligated with any blunt end. A linear DNA molecule may also be circularized using DNA ligase.

The following protocol may be used for DNA ligase reaction. If supplier supplies ready-made reaction buffer, then ready-made reaction buffer supplied by the supplier should be preferred:

- 66 mM Tris-HCl, pH 7.6
- 6.6 mM $MgCl_2$
- 10 mM DTT
- 0.4 mM ATP
- DNA ligase (few units, say nearly 10 to 20)

The incubation may be done either at 12°C for 16-24 hours or at 23°C for 5 hours. Reaction may be stopped by incubating in a water bath at 65°C for 10 minutes. Thereafter, if desired, phenol-chloroform extraction may be carried out followed by precipitation of DNA by alcohol for removal of protein. However, Scientists prefer to use the reaction mixture directly for transformation (without treating with phenol-chloroform).

References

[1] Lillehaug, J.R. and Kleppe, K. (1975) Effect of salts and polyamines on T4 polynucleotide kinase. *Biochemistry* **14:** 1225-1229.

[2] Panet, A, Van de Sande, J.H., Loewen, P.C., Khorana, H.G., Raae, A.J., Lillehaug, J.R. and Kleppe, K. (1973) Physical characterization and simultaneous purification of bacteriophage T4 induced polynucleotide kinase, polynucleotide ligase and deoxyribonucleic acid polymerase. *Biochemistry* **12:** 5045-5050.

[3] Maxam, A.M. and Gilbert, W. (1980) Sequencing end-labeled DNA with base specific chemical cleavage. *Methods Enzymol.* **65:** 499-560.

[4] Kornberg, A. (1988) DNA replication. *J. Biol. Chem.* **263**: 1-4.

[5] Richardson, C. C., Schildkraut, C.L., Aposhian, H.V. and Kornberg, A. (1964) Enzymatic synthesis of deoxyribonucleic acid. XIV. Further purification and properties of deoxyribonucleic acid polymerase of *Escherichia coli*. *J. Biol. Chem.* **239**: 222-232.

[6] Meinkoth, J. and Wahl, G.M. (1987) Nick translation. *Methods Enzymol.* **152**: 399-407.

[7] Joyce, C.M. and Grindley, N.D.F. (1983) Construction of a plasmid that over produces the large proteolytic fragment (Klenow fragment) of DNA polymerase I of *Escherichis coli*. *Proc. Natl. Acad. Sci. USA* **80:** 1830-1834.

[8] Zoller, M.J. and Smith, M. (1983) Oligonucleotide directed mutagenesis of DNA fragments cloned into M13 vectors. *Methods Enzymol.* **100:** 468-500.

[9] Sanger, F., Nicklen, S. and Coulson, A,R. (1977) DNA sequencing with chain terminating inhibitors. *Proc. Natl. Acad. Sci. USA* **74:** 5463-5467.

[10] Gubler, U. (1987) *Methods Enzymol.* 154, 154.

[11] Chang, L.M.S. and Bollum, F.J. (1986) Molecular Biology of terminal transferase. *Critical Reviews Biochem.* **21:** 27-52.

[12] Sambrook, J., Fritsch, E.F. and Maniatis, T. (1989) in *Molecular Cloning: A Laboratory Manual*, Cold Spring Harbor Laboratory Press, New York.

[13] Roychouchury, R., Jay, E. and Wu, R. (1976) Terminal labeling and addition of homopolymer tracts to duplex DNA fragments by terminal deoxynucleotidyl transferase. *Nucleic Acids Res.* **3:** 863-877.

[14] Kossel, H. and Roychouchury, R. (1971) *Eur. J. Biochem.* 22, 271.

[15] Huynh, T.V., Young, R.A. and Davis, R.W. (1984) in *DNA Cloning Techniques: A Practical Aproach*, IRL Press, Oxford

[16] Helfman, D.M. et al. (1987) *Methods Enzymol.* 152,354.

[17] Wu,R. et al. (1987) *Methods Enzymol.* 152, 343.

[18] Pheiffer, B.H. and Zimmerman, S.B. (1983) Polymer stimulated ligation: Enhanced blunt – or cohesive end ligation of DNA or deoxyribonucleotides by T4 DNA ligase in polymer solution. *Nucleic Acids Res.* **11**: 7853-7871.

Chapter V

Gel Electrophoresis

5.1. Introduction

Gel electrophoresis is considered to be an important technique in molecular biology. It is a direct method of separation and visualization of nucleic acid fragments. Gel electrophoresis is not only used as an analytical tool but it is also routinely used preparatively for the separation and purification of DNA fragments. The method depends upon the principle that nucleic acids are polyanionic at neutral pH and carry multiple negative charges due to the phosphate groups on the phosphodiester back bone of the nucleic acid strands. The negatively charged nucleic acid molecules migrate towards the positive electrode (anode) when placed in an electric field. The techniques are carried out using a gel matrix, which separates the nucleic acid molecules according to size. Two types of gels are commonly used, viz. agarose and polyacrylamide gels. Agarose is convenient for separating DNA fragments ranging in size from a few hundred to about 20kb whereas polyacrylamide is preferred for smaller DNA fragments (less than one Kb) and also for DNA sequencing.

5.2. Basis of Separation

The separation of DNA molecules by molecular weight occurs during gel electrophoresis but the mechanism responsible for separation is not well understood. The migration of the DNA molecules through the pores of the agarose (matrix) plays an important role in separation of the molecules on the basis of molecular weight/ size since the electrophoretic mobility of DNA in free solution is independent of molecular weight/ size. An agarose gel is a complex network of polymeric molecules whose average pore size depends on the buffer composition and the type and concentration of agarose used.

Agarose is a linear polymer composed of alternating residues of D- and L- galactose joined by α-1,3 and α-1,4 glycosidic linkages. The L-galactose residue has an anhydro bridge between the three and six positions. Chains of agarose form helical fibers that aggregate into supercoiled structures with a radius of 20-30 nm and gelation of agarose results in three-dimensional mesh of channels whose diameters range from 50 nm to >200 nm. Agarose is not homogeneous and the average length of the polysaccharide chains varies from batch to batch and from manufacturer to manufacturer. In addition, lower grades of agarose may be contaminated with other polysaccharides, as well as salts and proteins. This variability may affect gelling and melting temperature of agarose as well as the quality of the DNA recovered from the gel. These potential problems can be overcome by using special grades of agarose that are screened for the presence of inhibitors and nucleases and produce minimal background fluorescence after staining with ethidium bromide.

DNA molecules display elastic behavior by stretching in the direction of electrical field applied and then contracting into dense balls. The larger the pore size of the gel, the greater the ball of DNA which can pass through and hence the larger the molecules which can be separated. Once the globular volume of the DNA molecule exceeds the pore size, the DNA molecules can only pass through by repetition. This leads to size – independent mobility and loss of separation. If very large DNA molecules (1000-2000 kb) are to be separated by electrophoresis, it is necessary to use pulsed electrical fields by the technique called pulse field electrophoresis.

The migration rates of DNA molecules are inversely proportional to the logarithms of the molecular weights. The plot of DNA fragment length or molecular weight against the reciprocal of mobility gives a straight line over a wider range than the semi- logarithmic plot. Gel electrophoresis is frequently performed with DNA fragments markers of known size which allow accurate size determination of an unknown DNA molecule by interpolation. A particular advantage of gel electrophoresis is that the DNA bands can be readily detected at high sensitivity. The bands of DNA in the gel are stained with the intercalating dye ethidium bromide and as little as 0.05 ug of DNA in individual band can be detected under ultraviolet light as visible florescence. Gel electrophoresis, in addition to resolving DNA fragments of different lengths can be used to separate different molecular configuration of a DNA molecule. Additionally, the technique is also used for investigating protein – nucleic acid interactions and example is the so- called *gel retardation or band shift assay*. The band shift assay is based on the observation that binding of a protein to DNA fragments usually leads to a reduction in electrophoretic mobility. The assay typically involves the addition of protein to linear double – stranded DNA fragments, separation of complex and visualization of naked DNA by gel electrophoresis.

5.3. Factors Affecting Migration of DNA Through Agarose Gels

Size of DNA: Molecules of double- stranded DNA migrate through gel matrix at rates that are inversely proportional to the $\log_{10}$ of the number of base pairs. Larger molecules migrate slowly than smaller molecules.

Agarose concentration: A linear DNA fragment of a given size migrates at different rates through gels containing different concentrations of agarose. There is a linear relationship between the logarithm of the electrophoretic mobility of the DNA (μ) and the gel concentration (i) that is described by the equation:

$$\text{Log } \mu = \log \mu_0 - K_r.i$$

Where μ_0 is the free electrophoretic mobility of DNA and k_r is the retardation coefficient, a constant related to the properties of the gel and the size and shape of the migrating molecules.

DNA conformation: Superhelical circular (form I), nicked circular (form II), and linear (form III) DNAs migrate through agarose gels at different rates. The relative mobilities of the three forms depend primarily on the concentration and type of agarose used to make the gel, but they are also influenced by the strength of the current applied, the ionic strength of the buffer, and the density of superhelical twists in the form I DNA. To distinguish between the different conformational forms of DNA is simply to include in the gel a sample of untreated circular DNA and a sample of same DNA that has been linearized by digestion with a restriction enzyme that cleaves the DNA at only one place.

The presence of ethidium bromide in the gel and electrophoresis buffer: Intercalation of ethidium bromide causes a decrease in the negative charge of the double-stranded DNA and an increase in both its stiffness and length. The rate of migration of the linear DNA-dye complex through gels is consequently retarded by a factor of ~15%.

Voltage applied: At low voltage, the rate of migration of linear DNA fragements is proportional to the voltage applied. However, as the strength of the electric field is increased, the mobility of high-molecular-weight fragment increases differentially. Thus, the effective range of separation in agarose gels decreases as the voltage increases. Agarose gels should be run at no more than 5-8 V/cm for maximum resolution of DNA fragments of >2kb in size.

As written above, nucleic acids have a consistent negative charge and that's why move towards the anode. A potential gradient (**E**) is generated when a potential difference (voltage) is applied across the two electrodes. The potential gradient (**E**) is defined as the applied voltage, **V**, divided by the distance, **d**, between the two electrodes. When a potential gradient is applied, the force on a nucleic acid molecule bearing a charge of **q** Coulombs equivalent to **E.q**. Newtons drives a charged molecule towards an electrode of opposite charge. During movement of nucleic acid molecules, frictional resistance retards the movement of the charged molecules. The frictional force is a measure of the hydrodynamic size and shape of the molecule, the pore size of the gel matrix and viscocity of the buffer.

The velocity, **v**, of a charged molecule in an electric field is therefore given by the equation:

$$v = E.q/ f$$

Where '**f**' is the frictional coefficient

The movement of the nucleic acids in the gel may also be explained by the following equation:

$$\text{Log } m = \log q/f - K.c$$

Where **m** is mobility of the nucleic acid molecule

q is charge on the nucleic acid molecule in Coulombs

f is frictional coefficient which is dependent on the hydrodynamic size and shape of the molecule, the pore size of the gel matrix and viscocity of the buffer

K is the agarose interaction factor which is specific for each nucleic acid molecule and is dependent on the size and shape of the molecule

c is net charge on the molecule

When a potential difference is applied, molecules with different overall charges begin to separate due to different electrophoretic mobilities. Even the molecules of identical charges separate if they have different size due to different frictional forces.

Quality of agarose used: Standard agarose and low melting temperature agarose are the two major classes of agarose, whereas, third type consists of intermediate melting/gelling temperature agarose, exhibiting properties of each of the two major classes. For specialized applications, within each class there are various types of agarose.

The electrophoresis buffer: The electrophoretic mobility of DNA is affected by the composition and ionic strength of the electrophoresis buffer. In the absence of ions (e.g., if water is substituted for electrophoresis buffer in the gel or in the reservoirs), electrical conductivity is minimal and DNA migrates slowly, if at all. In buffer of high ionic strength (e.g., if 10x electrophoresis buffer is mistakenly used), electrical conductance is very efficient and significant amount of heat is generated, even when moderate voltage is applied. In most cases, the gel melts and the DNA denatures.

5.4. Electrophoresis Buffers

Several different buffers are available for electrophoresis of native, double-stranded DNA and these are (a) Tris-acetate-EDTA, pH 8.0 (TAE) (b) Tris- borate-EDTA, pH 8.0 (TBE) and (c) Tris-phosphate-EDTA, pH 7.5-7.8 (TPE). For laboratory use, electrophoresis buffers are usually made up as concentrated solutions and stored at room temperature.

Table. Composition of Electrophoresis Buffers

Buffer	Working solution	Stock solution/litre
TAE	1x 40 mM Tris-acetate,1 mM EDTA	50x 242g of Tris base, 57.1ml of glacial acetic acid, 100 ml of 0.5 M EDTA (pH 8.0)
TPE	1x 90mM Tris-phosphate, 2 mM EDTA	10x 108g of Tris base, 15.5 ml of phosphoric acid (85%, 1.679 g/ml), 40 ml of 0.5 M EDTA (pH 8.0)
TBE	0.5x 45mM Tris-borate,1 mM EDTA	5x 54 g of Tris base, 27.5 g of boric acid, 20 ml of 0.5 M EDTA (pH 8.0)

TAE has the lowest buffering capacity and gets exhausted on prolonged electrophoresis and when this occurs the anodic portion of the gel becomes acidic and the bromophenol blue migrating through the gel towards the anode changes its color from bluish-purple to yellow.

Exhaustion of TAE can be avoided by periodic replacement of the buffer during electrophoresis. Both TBE and TPE are slightly more expensive than TAE, but they have significantly higher buffering capacity. Double-stranded linear DNA fragments migrate ~ 10% faster through TAE than through TBE or TPE. The resolving power of TAE is slightly better than TBE or TPE for high molecular weight DNAs and worse for low-molecular weight DNAs. TAE yields better resolution of complex DNA fragments such as mammalian DNA. Southern blots used to analyze complex genomes are generally derived from gels prepared on and run with TAE electrophoresis buffer. The resolution of supercoiled DNAs is better in TAE than in TBE.

These buffers are usually prepared and stored as a 5x or 10x stock solutions. The pH of the concentrated stock buffer should be ~8.0. Concentrated stock buffer may be diluted to 1x just before use to make the gel solution as well as the electrophoresis running buffer from the same concentrated stock solution. Some investigators prefer to use more concentrated stock solutions (10x as opposed to 5x). However, 5 x stock solutions are more stable because the solutes do not precipitate during storage. Concentrated buffer stocks filtered through a 0.45-mm filter can prevent or delay formation of precipitates.

5.5. Gel Loading Dye Buffers

Gel- loading buffers are mixed with the DNA samples before loading into the slots of the gel and these buffers increase the density of the sample, ensuring that the DNA sinks evenly into the well and they add color to the sample, thereby facilitating the loading process. They contain dyes bromophenol blue and xylene cyanol FF, that in an electric field, move towards the anode at predictable rates. Bromophenol blue migrates through agarose gels ~ 2.2-folds faster than xylene cyanol FF, independent of the agarose concentration. Bromophenol blue migrates through agarose gels run in 0.5x TBE at approximately the same rate as linear double- stranded DNA 300 bp in length, whereas xylene cyanol FF migrates at approximately the same rate as liner double-stranded DNA 4 kb in length. Nowadays, some people prefer to use only bromophenol blue instead of combined bromophenol blue and xylene cyanol FF. Generally, 6x sample loading dye is 0.25 % bromophenol blue, 0.25 % xylene cyanol FF, 30 % glycerol in aqueous solution.

5.6. Protocols

1. Seal the edges of the open ends of the plastic tray supplied with the electrophoresis apparatus with tape to form a gel mould. Set the mould on the horizontal section of the bench.
2. Prepare sufficient electrophoresis buffer (usually 1x TAE or 0.5 x TBE) to fill the electrophoresis tank as well as to cast the gel. The same batch of electrophoresis buffer in both the electrophoresis tank and the gel should be used.
3. Prepare a solution of agarose in electrophoresis buffer at a concentration appropriate for separating the particular DNA size fragments (see the following Table). Weigh

correct amount of powdered agarose in an Erlenmeyer flask and add a measured quantity of electrophoresis buffer.

AGAROSE CONCENTRATION IN GEL (%,w/v)	AVERAGE SIZE OF SEPARATION OF LINEAR DNA MOLECULES (kb)
0.3	5-60
0.6	1-20
0.7	0.8-10
0.8	0.5-7
1.2	0.4-6
1.5	0.2-3
2.0	0.1-2

Range of Separation of DNA in Gels Containing Different Amounts of Standard Low-EEO Agarose.

4. To dissolve agarose, heat the slurry for the minimum time required to allow all of the grains of agarose to dissolve completely. Undissolved agarose appears as small "lenses" or translucent chips floating in the solution. Check that the volume of the solution has not been decreased by evaporation during boiling otherwise replenish with water if necessary.
5. Allow the molten gel to cool up to 55° C, add ethidium bromide to a final concentration of 0.5 μg/ml. Mix the gel solution thoroughly by gentle swirling.
6. While the agarose solution is cooling, choose an appropriate comb for forming the sample slots in the gel. Position the comb 0.5-1.0 mm above the plate so that a complete well is formed when the agarose is added to the mould.
7. Pour the warm agarose solution into the mould and ensure that no air bubble is entrapped in the gel. Air bubbles present in the molten gel can be removed easily by poking them with the corner of a Kimwipe. The gel should be between 3 mm and 5 mm thick.
8. Allow the gel to set completely (30-45 minutes at room temperature), then pour a small amount of electrophoresis buffer on top of the gel, and carefully remove the comb. Pour off the electrophoresis buffer and carefully remove the tape. Mount the gel in the electrophoresis tank.
9. Electrophoresis buffer in the tank should be just enough to cover the gel to a depth of ~ 1mm. It is not necessary to pre-run an agarose gel before the samples are loaded.
10. Mix the DNA sample with 0.20 volume of the desired 6x gel-loading dye buffer.
11. Slowly load the sample mixture into the slots of the submerged gel using a disposable micropipette, load DNA size standards (markers) into slots on both the right and left sides of the gel.
12. Close the lid of the gel tank partially and attach the electrical leads so that the DNA will migrate toward the positive anode (red lead). Apply a voltage of 1-5 V/cm (measured as the distance between the positive and negative electrodes). If the leads have been attached correctly, bubbles should be generated at the anode and cathode (due to electrolysis), and within a few minutes, the bromophenol blue should migrate from the wells into the body of the gel. Run the gel until the bromophenol blue and

xylene cyanol FF have migrated at an appropriate distance through the gel generally up to the bottom end of the gel.

13. Following the run turn off the electric current and remove the leads and lid from the gel tank. If ethidium bromide is present in the gel, examine the gel under UV trans-illuminator. Otherwise, stain the gel by immersing it in electrophoresis buffer or water containing ethidium bromide (0.5 μg/ml) for 5-10 minutes at room temperature or by soaking in a 1: 10,000-fold dilution of SYBER Gold stock solution in electrophoresis buffer.
14. Photograph the gel under UV transilliuminator as described (See Photography of gel).

5.7. Detection of Nucleic Acids

Staining and visualization under ultraviolet at 300 nm may detect DNA or RNA that has been subjected to electrophoresis through agarose gels by staining with ethidium bromide or SYBER Gold.

(A) Ethidium Bromide Staining of Nucleic Acids in Gels

Most commonly used dye to visualize DNA in agarose gel is a fluorescent dye ethidium bromide which contains a trycyclic planar group that intercalates between the stacked bases of DNA and binds to DNA with little or no sequence preference. Ultraviolet radiation at 254 nm is absorbed by the DNA and transmitted to the dye; radiation at 302 nm and 360 nm is absorbed by the bound dye itself. In both the cases, the energy is re-emitted at 590nm in the red orange region of the spectrum. The ethidium bromide –DNA complex has fluorescence – 20-30-fold greater than that of unbound dye and bands containing 10 ng of DNA can be detected in the presence of ethidium bromide (0.5 μg/ml) in the gel. Ethidium bromide is prepared as a stock solution of 10mg/ml in water stored at room temperature in Doric bottles wrapped in aluminum foil. In agarose gels ethidium bromide (one μl ethidium bromide dye solution per 10 ml gel solution) can be added in the gels so that gels can be examined directly under UV transilluminator during or at the end of the run. However, sharper bands of DNA are obtained when electrophoresis is carried out in the absence of the ethidium bromide and stained with ethidium bromide after the run. Ethidium bromide can not be added in polyacrylamide gels while casting, because it inhibits polymerization of the acrylamide, hence acrylamide gels are stained with ethidium bromide after the gel has been run.

(B) Syber Gold Staining of DNA in Gels

SYBER Gold is proprietary name of fluorescent asymmetrical cyanide dye used for staining single and double stranded nucleic acids in gels. This is marketed by Molecular Probes, USA. Syber gold has several advantages over ethidium bromide, as it is more sensitive and can be used to stain both DNA and RNA in the polyacrylamide and agarose gels containing denaturants such as urea, ethylene glycol and formaldehyde. Under standard UV transilluminator at 300 nm, SYBER Gold gives bright gold fluorescence that can be captured on conventional black and white film. The background fluorescence is so low that no destaining is required. But, this dye can not be added to gel before electrophoresis as its presence in the gel will cause severe distractions in the electrophoretic properties of DNA and RNA.

SYBER Gold may detect <20 pg of double stranded DNA, 100 pg of single standard DNA and 300 pg of RNA. SYBER Gold shows maximum excitation at 495 nm and has a secondary excitation peak at 300nm. Fluorescent emission occurs at 537nm. Following electrophoresis, SYBER Gold is used to stain DNA by soaking the gel, in 1: 1000 fold dilution of stock dye solution.

5.8. Pulsed Field Gel Electrophoresis

In 1984, Schwartz and Contor described pulsed field gel electrophoresis (PFGE) to resolve extremely large DNA molecules, raising the upper size limit of DNA separation in agarose from 30-50 Kb to over 10 Mb (10,000 Kb). After this initial report, many research papers described new and improved instrumentation and methods. Nowadays, several commercial pulsed field units are available. Now with pulsed field gel electrophoresis, larger size DNA molecules can be separated, cloned and analyzed instead of cloning smaller size fragments. It is pertinent to mention that during continuous field electrophoresis, DNA above 30-50 Kb migrates with the same mobility regardless of size. This is seen in a gel as a single large diffuse band. However, if the DNA is forced to change direction during electrophoresis, different sized fragments with in this diffuse band begin to separate from each other. With each reorientation of the electric field relative to the gel, smaller sized DNA begin to move in the new direction more quickly than the larger sized DNA. Therefore, the larger sized DNA lags behind providing a separation from the smaller sized DNA.

Although many types of PFGE equipments are available, they generally fall into two categories. The simplest equipment is designed for field inversion gel electrophoresis. Field inversion gel electrophoresis works by periodically inverting the polarity of the electrodes during electrophoresis. Because field inversion gel electrophoresis (FIGE) subjects DNA to a 180° reorientation, the DNA spends a certain amount of time moving backwards. Only an electrical field module is needed, and any standard vertical or horizontal gel box that has temperature control can be used to run the gel.

The other category contains equipments that reorient the DNA at smaller oblique angle, generally between 96 and 120°. This causes DNA to always move forward in a zigzag pattern down the gel. For a similar size range under optimal conditions, these separations are faster, resolve a wider size range, and give a larger useful portion of the gel compared to field inversion gel electrophoresis.

Ideally, the DNA should separate in straight lanes to simplify lane to lane comparisons. The original pulsed field systems used in homogeneous electric fields that did not produce straight lanes, making interpretation of gels difficult. The simplest approach to straight lanes is field inversion gel electrophoresis (FIGE) which uses parallel electrodes to assure a homogeneous electric field. The FIGE is a good technique using reorientation angle of 180°, and is able to separate DNA up to 1000 Kb. However, its separation range is considered to be under 2000 Kb. It has mobility inversions in which larger sized DNA can move ahead of smaller sized DNA during electrophoresis.

Ramping, where the reorientation pulse length is constantly increased during separation, will minimize inversions. This capability is included in most commercial equipments. Increasing both the separation range and the resolution of large sized DNA requires smaller reorientation angles in the range of 96 to 140°. Smaller angles increase the mobility of the DNA without much affecting resolution. However, the lower limit of reorientation angle is 96°, below this value, separation is not good.

Rotating gel electrophoresis is commercial introduction of pulsed field gel electrophoresis. It combines variable angles with a homogeneous electric field. The electrodes are positioned along opposite sides of the buffer chamber with their polarity fixed. The gel is cast on a circular running plate and afterwards placed in the buffer chamber. The gel is coupled to a magnetic drive beneath the buffer chamber to eliminate the possibility of leakage. To force the migrating DNA to a new direction, the magnetic drive simply rotates the gel to the new angle. The reorientation angle of the DNA is determined by a straight forward mechanical coupling. It offers much flexibility and voltage, angle and pulse times can be varied with the program stored in the memory of the unit.

It is difficult to isolate large sized DNA due to shearing problem and it is also difficult to pipet solution of large sized DNA due to high viscosity. Therefore, to avoid the problem, bacteria or yeast is first embedded in agarose plugs and afterwards, the plugs are treated with enzymes to break the cell wall and proteins leaving undamaged naked DNA in the agarose. The plugs having naked DNA are cut to size, treated with restriction enzymes if necessary and loaded in the sample well and sealed into place with agarose.

Generally, in pulsed field gel electrophoresis, voltage 180 V (at 4-6 V/cm), pulse length 90 seconds, reorientation angle 120° and 0.5x TBE buffer is used. The electrophoresis is run in cold condition. Most systems separate DNA over a relatively small area limiting the resolution of complex samples. However, rotating gel electrophoresis uses more separation distance (nearly 20 cm) and gel size is normally 18 x 20 cm.

The field strength has much effect on separation and is a compromise between separation time and resolution of a particular size class. Four to six volts/ cm is generally required for resolving DNA up to 2000 Kb in a reasonable period of time (1-2 days). These field strengths trap and immobilize even bigger DNA in the agarose matrix and bigger DNA (more than 3000 Kb) requires 2V /cm or less for separation. The pulse time is also important since it changes the size range of separation. Longer pulse times lead to separation of larger DNA.

The commonly used buffers in pulse field gel electrophoresis are TAE (1x TAE is 40 mM tris-acetate, 1 mM EDTA, pH 8.0) and TBE (1 x TBE is 90 mM tris, 90 mM boric acid, 2 mM EDTA, pH 8.0). However, both the buffers are used at relatively lower ionic concentration to prevent heating. Generally, 0.25x or 0.5x buffers are used. In addition to prevent heating, low ionic buffers increase DNA mobility. It has been found that if 0.25x buffer is used, DNA mobility is 40-50% faster compared to when 1x buffer is used. However, one can not use lower than 0.25x concentration of the buffer since in that case, DNA mobility will drop off.

The quality of the agarose is also important for separation of large size DNA molecules. It has been found that with low electroendosmosis (EEO) agarose, DNA molecules move faster and there is better resolution. Nowadays, many companies sell special low EEO pulsed field grade agarose. The concentration of agarose also affects the separation and resolution of DNA molecules. It has been found that high concentrations of agarose yield sharper, but slower moving bands. The typical concentrations used (0.8 to 1.2 %) represent a compromise between speed and resolution.

DNA mobility also depends on the separation temperature. The temperature must be constant both during and between runs. Although higher temperatures increase DNA mobility but there is poor resolution. Many people prefer to perform electrophoresis at nearly 10°C.

5.9. Polyacrylamide Gel Electrophoresis for Nucleic Acids

For separation and purification of smaller nucleic acids fragments (less than one Kb size), native polyacrylamide gel electrophoresis is used. The basis of separation in polyacrylamide gel is also the same as in agarose gel. However, due to small controlled pore size in polyacrylamide gel, separation of larger-sized DNA fragments do not take place. Generally for the separation of nucleic acid molecules, 5 to 7% polyacrylamide gel is used (larger the size of the nucleic acid molecules, the lesser the percentage of the polyacrylamide gel). In polyacrylamide gel electrophoresis, the gel is prepared by polymerizing acrylamide and N, N'- methylene bisacrylamide which form a network of pores, the sizes of which are controlled by the initial concentration of acrylamide and methylene bisacrylamide. Normally, for electrophoresis of nucleic acids, ratio of acrylamide and methylene bisacrylamide is taken 19:1, whereas for proteins, the ratio is taken 29.2: 0.8. Some people prefer to create a gradient of concentration of the monomers and hence a variation in pore size going from the top of the gel to the bottom. Here also the same buffers viz. TAE, TPE, TBE as in case of agarose gel electrophoresis are used as gel and electrode buffers. A typical protocol for the electrophoresis of nucleic acids is as follows:

For 100 Ml Solution (6% Gel)

- 5.7 g Acrylamide
- 0.3 g Methylene bisacrylamide
- 10 ml 10x electrode buffer
- 10 μl tetramethylethylenediamide (TEMED)

Dissolve in 60-70 ml of water using heat (heating for 60 seconds in a microwave oven), make the volume 100 ml with water

De-gas the solution using a suction pump

- 70 mg Ammonium persulfate

Dissolve rapidly and pour in a ready cascade of the glass plates, insert suitable comb for making wells

Allow to gellify for about 30-40 minutes

Afterwards, the gel along with glass plates is mounted in the electrophoresis chamber as per instructions of the supplier. After applying the sample of the nucleic acid mixed with the loading dye (sample buffer), the electric current is applied and the sample is electrophoresed till the bromophenol blue dye reaches at another end of the gel. Generally 100 to 150 mA current is applied. However, amount of the current to be applied depends on the length and thickness of the gel. After stopping the electrophoresis, the gel is removed carefully from the glass plate and put in a tray containing water with few drops of ethidium bromide dye. After 5-10 minutes, the DNA in the gel is visualized using UV trans-illuminator in the same way as in agarose gel electrophoresis. As written above, SYBER gold staining may also be done in place of ethidium bromide.

5.10. Polyacrylamide Gel Electrophoresis for Proteins

For electrophoresis of proteins under non-denaturing conditions, the basic procedure is same except the ratio of acrylamide and methylene bis acrylamide is taken 29.2: 0.8. The electrode buffer used is generally 0.05M tris-HCl, pH between 7.5 and 8.5. Some people have also used other buffers like phosphate buffer, tris-glycine buffer. Many people prefer to prepare separating gel (small pore size gel) and stacking gel (larger pore size gel) as first described by Ornstein in 1964. The stacking gel generally has 3% polyacrylamide gel which acts to compact the protein bands (without separation) whereas separating gel acts to separate the protein bands on the basis of charge as well as size of the proteins. Nowadays, sodium dodecyl sulfate (SDS) polyacrylamide gel electrophoresis is more in use for proteins rather than non-denaturing electrophoresis. Sodium dodecyl sulfate, an anionic detergent, is used to completely suppress the native charge on the proteins. It gives them a large negative coat of detergent molecules. The SDS also interacts with the hydrophobic core of proteins, it causes a rapid and irreversible unfolding that linearizes the polypeptide chains which all now have a roughly equivalent charge/ mass ratio. Therefore, in SDS polyacrylamide gel electrophoresis, the separation of proteins is on the basis of molecular weight (size) of the proteins. Nowadays, it is considered to be a very useful and high resolution analytical method. The SDS polyacrylamide gel electrophoresis is of little importance for the purification of proteins. However, it is routinely used throughout the purification of proteins to test the efficiency of the purification. The SDS polyacrylamide gel electrophoresis is also used for determination of subunit molecular weight of proteins using standard protein markers. A plot of log molecular weight versus migration distance (protein mobility) on the gel should be linear. In the laboratory, for SDS polyacrylamide gel electrophoresis, SDS is used in the gel solution as well as in the electrode buffer. Nowadays, ready-made commercial gels are also available which are stored in $-20^{o}C$ freezer before use.

For SDS polyacrylamide gel electrophoresis, sample of the proteins is given treatment so that proteins remain in denatured, unfolding condition. Many sample buffers are in use for the purpose and all work on the same principle to denature proteins. A commonly used 2x sample buffer is 2% SDS, 20% glycerol, 20 mM tris-HCl, pH 8, 2 mM ethylene diamine tetra acetate (EDTA), 80 mM dithiothreitol (DTT) and 0.2% bromophenol blue to serve as tracking dye. The protein concentration of the unknown sample is usually adjusted to 2 mg/ml and equal amount of the sample buffer is added. In the sample buffer, EDTA is a preservative that chelates bivalent cations which reduces the activity of proteolytic enzymes that require calcium and magnesium ions as cofactors. The tris acts as a buffer which is very important since the stacking process in discontinuous electrophoresis requires a specific pH. The glycerol makes the sample more dense than the electrode buffer, so the sample will remain in the bottom of a well rather than float out. The bromophenol blue dye helps in tracking the progress of the electrophoresis. Sodium dodecyl sulfate, dithiothreitol and heat are responsible for the actual denaturation of the sample. Sodium dodecyl sulfate breaks up the two and three dimensional structure of the proteins by adding negative charge to the amino acids. The same charges repel, therefore, the proteins are more or less straightened out rendering them functionless. Some quarternary structures may remain due to disulfide bonding and other covalent and non-covalent bonds to other types of molecules.

A typical protocol for slab gel discontinuous SDS polyacrylamide gel electrophoresis is as follows:

Reagents

Acrylamide (30% stock)

- 29.2 g Acrylamide
- 0.8 g Methylene bis acrylamide

Dissolve in water and make the volume 100 ml with water.

Lower gel stock solution (1M tris, pH 8.8, 0.25% SDS)

- 60.6 g Tris
- 1.25 g SDS

Dissolve in water, adjust the pH 8.8 with HCl and make the volume 500 ml with water.

Upper gel stock solution (0.5M tris, pH 6.8, 0.25% SDS)

- 15.2 g Tris
- 0.63 g SDS

Dissolve in water, adjust the pH 6.8 with HCl and make the volume 250 ml with water.

Running electrode buffer (0.19 M glycine, 25 mM tris, pH 8.3; 0.1 % SDS)

- 7.2 g Glycine
- 1.5 g Tris
- 0.5 g SDS

Dissolve in water, adjust the pH 8.3 and make the volume 500 ml with water.

SDS Cocktail

- 1 ml 50% glycerol
- 1 ml Upper gel stock solution

- 100 µl 2- mercaptoethanol
- 200 µl Bromophenol blue (2 % in 40% sucrose and 0.2% SDS)

Staining solution

Dissolve 1.25 g Coomassie brilliant blue R-250 in 227 ml of ethanol using hot magnetic stirrer. Afterwards add 46 ml acetic acid and followed by 227 ml of water.

Stir, filter through Whatman No. 1 filter paper.

De-stain solution

- 667 ml water, 250 ml ethanol and 83 ml acetic acid

For Making 10% Polyacrylamide Gel (Separating Gel)

- Lower gel stock solution 11 ml
- Acrylamide 10 ml
- Water 8.8 ml
- TEMED 19 µl
- 10% SDS solution 300 µl

De-gas using a suction pump.

Add 150 µl 10% ammonium persulfate solution, shake well and pour the solution carefully in the glass plates cascade.

For Making 3 % Stacking Gel

- Upper gel stock solution 5 ml
- Acrylamide 3 ml
- Water 11.8 ml
- TEMED 28 µl

De-gas using a suction pump.

Add 200 µl 10% ammonium persulfate solution, shake well and pour the solution carefully in the glass plates cascade above the separating gel (after separating gel gets solidified). Pour little water above the gel solution in order to avoid drying of the gel.

Once the gel solution get solidified, remove the comb and place the gel in the electrophoresis chamber filled with the electrode buffer, apply sample and run electrophoresis using a suitable current. The current should be stopped when the bromophenol blue tracking dye reaches at the another end.

After electrophoresis, the gel may be placed in staining solution for about 30 minutes and afterwards de-stained using de-stain solution. The protein bands will be visualized as purple colored bands due to binding of Coomasie brilliant blue dye.

Isoelectrofocusing

Some people prefer to use isoelectrofocusing as an analytical tool for checking the purity of the proteins. In isoelectrofocusing, a pH gradient is generated in the gel by using suitable range ampholyte. In isoelectrofocusing, the separation is on the basis of isoelectric pH (IpH) of the proteins. It is considered to be a powerful analytical tool where proteins even having a IpH difference of 0.01 can be separated. Not only this, it is said that if Immobiline is used, the proteins having a difference of 0.001 in IpH may also be separated. That's way, isoelectrofocusing is a powerful analytical tool.

A typical protocol for isoelectrofocusing of the proteins is as follows:

Preparation of gel:

- 5.3 ml water
- 2 ml 50% glycerol
- 2 ml 22.2 % acrylamide containing 0.6 % methylene bis acrylamide
- 3 μl TEMED
- 0.64 ml 40 % (W/V) Ampholine (carrier ampholyte) of desired pH range (it should correspond to an ampholine concentration about 2.5 % in the gel)

De-gas the mixture using suction pump, thereafter, add a few crystals (5 mg) of solid ammonium persulfate, dissolve and immediately transfer in the gel tube/ plate and allow it to polymerize which takes about 30 minutes. The 1 M orthophosphoric acid solution may be used as anolyte solution and 1 M sodium hydroxide as catholyte solution. After polymerization of the gel, it may be placed in the electrophoresis chamber. Some people prefer to use lower volage like 400 V with longer duration of the electrophoresis while others prefer to apply higher voltage like 1400 V and separation time about 2 hours. After isoelectrofocusing run, the gel is placed in the fixing solution (17.25 g sulfosalicylic acid and 57.5 g trichloroacetic acid (TCA) in 150 ml of methanol and 350 ml of water) for one hour. Thereafter, the gel is stained for 10 minutes at 60 °C with coomasie brilliant blue dye (0.115 g dye per 100 ml of solvent consisting of 500 ml ethanol and 160 ml acetic acid made to 2 litre with water) and destained in dye free solvent at room temperature.

Some people even prefer to perform two dimensional (2D) polyacrylamide gel electrophoresis. In the technique, the methods of isoelectrofocusing and SDS polyacrylamide gel electrophoresis are combined to get a high resolution of the protein mixture. In fact, 2D polyacrylamide gel electrophoresis is considered to be the ultimate test of purity. Here, an isoelectrofocusing gel is prepared and run in a thin tube that matches the width of the SDS polyacrylamide gel. First, protein is applied on the isoelecrofocusing gel and run in an electric current, once focusing is complete, the isoelectrofocusing tube gel is laid sideways on top of the SDS polyacrylamide gel and an SDS containing buffer and current is applied. The proteins that were initially separated according to their isoelectric pH, are now separated according to their molecular weight.

5.11. Extraction of Nucleic Acid from Agarose and Polyacrylamide Gels

After electrophoresis and staining with ethidium bromide, nucleic acid band(s) may be visualized under UV trans-illuminator. One should always take care while handling ultraviolet irradiation. If the skin of the body gets exposed to the ultraviolet irradiation, there are chances of developing skin cancer due to mutation in the nucleic acids of skin cells. One should always wear goggle on the eyes, gloves on the hands, laboratory coat on the body in order to avoid exposure of UV irradiation. Once, bands of nucleic acid are visible under UV trans-illuminator, the desired band may be cut with the help of a razor blade or scalpel. While cutting the nucleic acid band, efforts should be done so that minimum amount of the gel should be cut without losing nucleic acid. The cut band is removed from the gel with the help of spatula (without touching by finger). With agarose gel as well as polyacrylamide gel, generally from the band, nucleic acid is extracted by electro-elution. For the purpose, the cut band of the gel (agarose or polyacrylamide) is placed in a dialysis tubing with 0.5 or 1 ml of the elution buffer. Generally, tris-acetate-EDTA (40 mM tris, 1 mM EDTA and 20 mM acetic acid, pH 8.3) is used as elution buffer. The dialysis tubing is sealed with the help of a clip and put in to a submerged type electrophoresis chamber. The elution buffer is also put in the electrophoresis chamber so that dialysis tubing should remain submerged in the buffer. The electrophoresis chamber is connected to the power supply with the help of electrode leads. The power supply is connected to the power line and switched on. Generally, 100 milli-ampere current is applied for about an hour. After an hour, the power supply is switched off and electrophoresis chamber can be carried in a dark room to check progress of the elution. For the purpose, ultraviolet rays are irradiated on the gel present with in the dialysis tubing with the help of a ultraviolet torch. If the gel still has nucleic acid, it will show fluorescence. If it is so, elution should be continued for another nearly 30 minutes (after connecting with the power supply). After elution, nucleic acid will be in the elution buffer put with in the dialysis tubing. Now, this buffer from the dialysis tubing is taken out carefully with the help of a glass pasteur pipet and transferred in an eppendorf tube. The buffer containing nucleic acid will be contaminated with the particles of the gel. Now, phenol-chloroform treatment is done to remove gel traces (as described in Chapter 1) followed by di ethyl ether treatment. Afterwards, nucleic acid is precipitated out using salt (sodium chloride or sodium acetate) and ethanol as described in Chapter 1. Finally, nucleic acid may be dissolved in water or TE buffer. Commercially, electro-elution chambers are also available which also work on the same principle. On passing electric current, the nucleic acid gets moved from the gel and gets collected in the buffer present at the anodic end. Nowadays, commercial kits are also available which are used for extraction of nucleic acid from agarose gel without carrying electro-elution. One such kit is available from QIAGEN Inc. The QIAGEN kit is designed by taking the use of spin-column technology with the selective binding properties of an uniquely designed silica gel membrane. The kit contains a buffer called as QG buffer which solubilizes the agarose gel slice and provides the appropriate conditions for binding of DNA to the silica membrane. DNA absorps to the silica membrane present in the spin column in the presence of high salt while contaminants pass through the membrane. Impurities are efficiently washed away, and the DNA is eluted with water or TE buffer. Adsorption of DNA to silica membrane also depends on pH. Adsorption is typically

95% if the pH is nearly 7.5 and is reduced drastically at higher pH. If the loading mixture pH is more than 7.5, the optimal pH for DNA binding can be obtained by adding a small amount of 3 M sodium acetate, pH 5.0. The QG buffer in the kit also contains a pH indicator, allowing easy determination of the optimal pH for DNA binding. The pH indicator in buffer QG appears yellow in the pH range 7.5. It becomes orange or violet if the pH becomes more than 7.5. If so, small amount of 3 M sodium acetate, pH 5.0 may be added before proceeding with the protocol. In addition, the color of the binding mixture allows easy visualization of any unsolubilized agarose, ensuring complete solubilization and maximum yields. As per claim by the manufacturer, the indicator dye does not interfere with DNA binding and is completely removed during the clean up procedure. As per claim of the manufacturer, buffer QG does not contain sodium iodide. In fact, residual sodium iodide may be difficult to remove from DNA samples which reduces the efficiency of subsequent enzymatic reactions. If one uses commercial kit for extraction of DNA from agarose gel, the instructions of the manufacturer should be followed.

5.12. Photography of Gels

Photography of ethidium bromide stained gels may be done using transmitted or incident UV light and most commercially available devices called trans-illuminators emit UV light at 302nm. The fluorescence of ethidium bromide DNA complex is considerably greater at this wavelength than at 366 nm and slightly less than at short wavelength (254 nm), however, the nicking of DNA is much less at 302 nm than 254nm. The photograph may be taken in the Polaroid film type 57 or 667 (ASA 3000) with an efficient UV light source as well as, a written 22A (red / orange filter) and a good lens (f =135 mm) with an exposure of few seconds.

Nowadays, integrated systems containing light source, fixed – focus digital cameras and thermal printers are available to capture, images, of ethidium bromide – stained gels. The CCD cameras of these systems use a wide – angle zoom lens (f = 75 mm) that allow detection of very small amounts of ethidium – bromide stained DNA (0.01–0.5 μg). In more advanced systems, gel images are directly transmitted to a computer and utilized in real time.

5.13. Minigel Electrophoresis

Minigels are particularly useful when a rapid information is required before the next step in a cloning protocol can be undertaken. Because the wells are smaller and the gels are thinner, less quantity of DNA is required for visualization. In addition, minigels require smaller amount of reagents and run rapidly, hence offer considerable saving in both time and money. Many investigators prepare one gel at the beginning of the week, and use it over and over again through the course of an experiment. Thus, a particular set of samples may be loaded, run out of the gel, and visualized. The gel then may be "erased" by running the sample off the gel into the buffer. Minigels are best suited for the analysis of small DNA fragments (<3 kb). Larger fragments resolve poorly because of the high voltages that are generally used and the comparatively short length of the gel. Several types of miniature electrophoresis tanks are manufactured commercially. Each gel slot holds 3-12 μl of sample, depending on the thickness of the gel and the width of the teeth of the comb. Usually, 10-100 ng of DNA in the gel-loading buffer of choice is applied to a slot. The gel is then run for 30-60 minutes at high voltage (5-20V/cm) until the bromphenol blue and xylene cyanol FF have migrated the appropriate distance. The gel is then photographed.

5.14. Polyacrylamide Gel Electrophoresis for DNA Sequencing

Analysis of DNA sequencing is performed on denaturing polyacrylamide gels. The gel contains urea for denaturation of DNA fragments. The samples of sequencing reactions are electrophoresed at high voltages so that heat also acts as a denaturant. Ultra-thin gels (0.2 mm) are used to increase the resolution of sequence bands. The length of the gel is 40 cm or 80 cm (some people even used 120 cm long gel) to maximize the number of nucleotides that can be read.

Various percentages of polyacrylamide may be used for sequence analysis. An 8% polyacrylamide gel is good for reading sequences in the range of 100 nucleotides. Whereas, 6 % gel with 80 cm length and multiple loading is found good for reading sequences in the range of 350 nucleotides. In sequencing gel electrophoresis too, tracking dyes, bromophenol blue and xylene cyanol FF are used. However, dyes solution is prepared in deionized formamide (0.25 % bromophenol blue, 0.25% xylene cyanol FF in 100% de-ionized formamide). In fact, the dye solution is used to stop the sequencing reaction in case of dideoxy sequencing and to dissolve DNA after chemical modification reactions in case of Maxam and Gilbert method. Migration of the dyes is used as an indication of the location of the DNA fragments. On an 8 % polyacrylamide gel the bromophenol blue migrates at a position equivalent to 15 bases, and the xylene cyanol migrates at a position equivalent to 75 bases.

If the electrophoresis of an 8 % polyacrylamide gel is terminated with the bromophenol blue at the bottom of the 40 cm gel, the sequence from approximately 15 to 70 can be read from a single loading of the sample. In order to resolve larger fragments and read the DNA sequence for more than 70 bases, multiple loadings of the sequencing reactions are performed and instead of 40 cm gel, 80 cm polyacrylamide gel is used. The samples are loaded a second time and electrophoresis is continued until bromophenol blue dye front of second loading reaches the bottom of the gel. During this period, the xylene cyanol in the samples that were loaded first will have migrated to the bottom of the gel. Some people even prefer to load samples third time. The sequence is read starting from the set of reactions that were loaded last. When the spacing between the bands becomes very close near the top of the gel, the reading may be continued in the lanes loaded before that loading.

Some people prefer to use buffer gradient gels which provide better resolution of the DNA fragments. This gel technique uses a buffer concentration gradient with in the gel where the buffer concentration is higher at the bottom of the gel than at the top. The gradient affects the mobility of the DNA fragments such that the bands are spaced evenly rather than logarithmically, allowing more bases to be read from a single gel loading.

Another improvement in sequencing gel electrophoresis is the use of shark tooth comb, Instead of forming conventional wells, the teeth of the comb are pointed. The design of this comb virtually eliminates space between adjacent sample lanes and increases the number of samples that may be loaded on a single gel. The resulting sequence bands may be read further with less ambiguity.

Reagents required:

- Acrylamide
- Methylene bisacrylamide
- Urea (Ultrapure)
- TBE buffer (composition given above in this chapter)
- TEMED
- Ammonium persulfate
- Ethanol
- Siliconizing solution (1 % (V/V) dimethyl dichloro silane in chloroform or carbon tetra chloride)

Please note that all the reagents should be high quality (if possible molecular biology grade)

Please also note that acrylamide and methylene bisacrylamide are neurotoxins and are readily absorbed through the skin. Therefore, inhaling acrylamide dust should be avoided and throughout, gloves should be weared. Di-methyl di-chloro silane is carcinogenic, therefore adequate precautions should be taken while handling it.

Urea acts as a denaturant in sequencing gels. Ultra pure urea is required for good resolution of the DNA fragments. The concentration of urea in the gel is kept 50 % (8.3 M). As written above, sequencing gels are electrophoresed at high voltages to generate heat and to keep the DNA denatured. The surface temperature of the glass plates should be between 50 to 60°C during electrophoresis. High voltage power supplies are required to obtain these temperatures. Generally, 5 to 10 KVA power supply unit is used in sequencing gel electrophoresis.

A standard 6 % denaturing polyacrylamide sequencing gel may be formulated as follows:

For 100 ml formulation

- Acrylamide 5.7 g
- Methylene bisacrylamide 0.3 g
- Urea 50 g
- 10x TBE buffer 10 ml

Add approximately 40-45 ml of water. Dissolve by heating in a microwave oven for about 60 seconds and make the volume 100 ml with water.

Add 10 μl TEMED

De-gas the solution using a suction pump.

Add 70 mg ammonium persulfate.

Pour gel in the glass plates cascade taking care of no entrapment of air bubbles.

Preparation of Glass Plates

Clean glass plates thoroughly using good quality non-abrasive detergent. The detergent should not leave a soap residue. Do not scrub plates. No scratch should be on the glass plates. Rinse glass plates thoroughly with de-ionized water. Wipe dry using paper towel.

The surface of plates which will be in contact of polyacrylamide gel should be siliconized. For this, apply 1% solution of di-methyl di-chloro silane (in chloroform or carbon tetra chloride) on the plates with the help of paper towel in a fume hood. Leave the plates for about 15 minutes in the hood. Afterwards, rinse with water thoroughly without applying any detergent. Rinse with de-ionized water and wipe dry with paper towel.

Gel plates cascade is prepared using suitable comb as per instructions of the supplier. The plates are taped around the edges using a sealing tape.

Either a 50 ml syringe can be used to inject the gel solution or pour carefully from the beaker between the plates. The pouring of the gel formulation should be complete with in 10 minutes. If one needs more time, amount of TEMED and/or ammonium persulfate may be reduced accordingly. To pour the gel formulation in to the plates, hold the plates at about 45° angle so that the gel solution flows down along one of the side spacers. An uniform flow of the solution should be maintained in order to avoid entrapment of air bubbles. Even after taking all the care, if air bubbles occur in the plates, place plates vertically and work the bubbles to the top by rapping on the glass plates. If air bubbles are not removed, one has to demantle the cascade and the entire exercise must be repeated. Continue to pour the gel solution so that there is liquid extruding from between the plates when the plates are filled and placed horizontally.

Placing the Comb

Lay the plates on bench top at a slight angle (nearly 5 to 10°) so there is a reservoir of gel solution at the top. Thereafter, carefully insert comb in top of the gel. Insert the comb only to the top of the well teeth. Place clamp on top of glass plate and over comb to prevent a layer of gel forming between the comb and the plate. Thereafter, leave the plates in the horizontal position for polymerization. The polymerization takes about 30-40 minutes. Some people prefer to leave the plates overnight after polymerization. If one wishes to leave the plates overnight or even for few hours, one should leave the comb in place and wrap the top of the plates with plastic wrap. When gel has to be used for electrophoresis, one should remove the tape and clamps and carefully remove comb. Some people prefer to add a small amount of buffer in the comb area before it is removed in order to prevent the wells from collapsing. After the comb is removed, the wells should be rinsed with buffer in order to insure that clean, sharp interface is present at the bottom of the well. Afterwards, the glass plates cascade having polyacrylamide gel is placed in to the electrophoresis chamber and fixed with the help of clamps. The electrophoresis chamber (both cathodic and anodic ends) should have sufficient amount of buffer so that both ends of the plates should remain dipped in buffer.

If one has to insert the shark tooth comb, then after putting the plates cascade on a bench top at a slight angle (5 to 10°), one should insert the flat edge of the comb between the plates to a depth of 2-3 millimeter below the short plate. By keeping the depth to a minimum, the samples may be loaded with a mechanical pipet. Afterwards, a clamp should be placed on both sides of the cascade at the top of the small plate. To eliminate the leakage between the wells, the clamp should be moved inward as far as possible. This will force the plates together and the comb will act as the spacer. This clamping should be done only in the comb area. Afterwards, the plates may be left in the horizontal condition for polymerization. After polymerization of the gel, the tape and clamp can be removed and the comb should be slide out from between the plates carefully. The top of the gel must be rinse carefully with electrophoresis buffer to remove any un-polymerized acrylamide. Thereafter, the comb should be rinsed with water to clean it and place the comb between glass plates with the teeth down towards the gel. The comb should be inserted until it just makes contact with the surface of the gel. The teeth of the comb should not be allowed to pierce the gel surface. A slight indentation of the gel surface will be evident when the comb has been properly inserted. One should not slide the comb laterally after it has come in contact with the acrylamide. Thereafter, the gel cascade may be clamped in to the electrophoresis chamber and upper and lower reservoir of the chamber should be filled with electrophoresis buffer.

Buffer Gradient Gels

Here, the gel contains a concentration gradient of the standard TBE buffer. The gradient alters the band spacing and permits more sequence to be read from a single sample loading. The buffer gradient gels allow an average of 200 nucleotides of sequence to be read from one sample loading. A typical protocol for buffer gradient gel is as follows:

Solution I (for 100 ml solution)

- Acrylamide 5.7 g
- Methylene bisacrylamide 0.3 g
- 10x TBE 5 ml
- Urea 50 g

Dissolve in water and make the volume 100 ml.

Solution II (for 100 ml solution)

- Acrylamide 5.7 g
- Methylene bisacrylamide 0.3 g
- 10x TBE 25 ml
- Urea 50 g
- 40% sucrose 25 ml
- Bromophenol blue 5 mg

Dissolve in water and make the volume 100 ml.

Pouring Buffer Gradient Gels

One should prepare the gel plates as described above. The plates are assembled with the side spacers and tape the edges with the sealing tape.

Transfer 65 ml of solution I and 12 ml of solution II in separate beakers. To solution I, add 300 μl of 10% ammonium persulfate and 70 μl of TEMED, and to solution II, add 60 μl of 10% ammonium persulfate and 20 μl of TEMED. Mix thoroughly but rapidly.

Immediately draw up in a 25 ml pipet of 10 ml of solution I followed by 10 ml of solution II carefully so there is as little mixing as possible.

Introduce the solution in to gel cascade by flowing it along one side spacer.

Add the remaining gel solution I to fill the gel plates cascade. Move the syringe along the top of the gel as the solution is introduced so the bottom layer is not displaced. The bromophenol blue dye allows one to monitor this process. However, an absolutely smooth and even gradient is not required to achieve good resolution of the sequencing bands.

Insert the comb in the same way as described above and allow to polymerize the gel.

Electrophorese the gel in 0.5x TBE at the appropriate voltage. The tracking dyes move very quickly at first but slow down dramatically as they move near the bottom due to high ionic strength. It will take somewhat longer to move the bromophenol blue dye to the bottom than on a standard gel.

Electrophoresis of Sequencing Gels

Gels should be pre-electrophoresed for 30-40 minutes before loading the samples. The electrophoresis is carried out at constant temperature (between 50 to 60°C). This is achieved by application of constant power. Alternatively, the gel can be electrophoresed on a constant voltage setting. Under these conditions, the amperage will decrease during electrophoresis. If the glass plates surface temperature becomes too high, the plates will crack.

Sample Application

After pre-electrophoresis, the gel is ready to be loaded. The DNA samples are heated at 90-100°C for 3 to 5 minutes just prior to loading and then transferred to ice. The sample to be electrophoresed should be applied only when the surface temperature of the glass plates is between 50 to 60°C. Immediately prior to loading of sample, the wells of the gel must be rinsed out using a pasteur pipet to remove urea that has diffused in to the well. If this is not done, the sample will not form a tight band and band sharpness on the autoradiogram will suffer. Either drawn out glass capillaries or ultra thin pointed pipet tips containing auto-pipet are used to load the samples in the conventional well combs. Generally, 1-3 μl of sample per well is applied. To prevent diffusion of the samples, not more than two sets of samples must be applied at a time, electrophorese for five minutes and then additional sets may be loaded.

With shark tooth combs, samples may be applied using a drawn out capillary pipet as is customarily done with conventional combs. A mechanical pipettor also may be used. Sample application with a pipettor produces satisfactory results. After drawing the sample in the tip, the same should be positioned above the sample well to dispense the sample.

Autoradiography

Thin sequencing gels can not be handled without some backing otherwise they will stretch and tear. Autoradiography can be done by leaving the gel on one of the plates or by transferring the gel to a suitable backing (Whatman filter paper No.3). If the gel is more than 10% polyacrylamide, X-ray film may be used as backing. To transfer the gel, remove one glass plate from the gel sandwich by prying the plates apart with a spatula. Press the backing firmly on to the gel, especially at the edges. Gently peel away the backing and the gel. If the gel does not stick well to the backing, turn the sandwich over and peel the glass plate off of the gel and backing. Thereafter, cover the gel with saran wrap (plastic wrap). If the gel is transferred on Whatman filter paper No.3, it may be dried under vacuum using nearly 60-65°C temperature in a gel drier. The gel is exposed to X-ray film for autoradiography of radioactive phosphorus samples. For sequencing reactions with ^{35}S-nucleotides, it may be necessary to fix the gel in 5% methanol/ 5% acetic acid for 15-20 minutes before drying. Due to the low energy of ^{35}S-emission, the gel should not be covered with plastic wrap during autoradiography.

Autoradiography is performed in metal, cardboard or homemade cassettes. Metal cassettes have a spring loaded lid which holds the gel and film in close contact for better autoradiography. Irrespective, what ever cassette is used, close contact between the film and the gel is necessary for sharp bands on the autoradiogram. Gels may be autoradiographed while frozen (-20 or -80°C) or at room temperature after drying. Dried gels yield sharper bands. If the gel is frozen after autoradiography, it should be brought to room temperature and wiped dry before a re-exposure is done to prevent water condensation on the new film. Some people use intensifying screens for quicker autoradiography. However, use of intensifying screens increases band broadening and decreases resolution and the ability to read the sequence on the upper part of the autoradiogram.

References

[1] Aaij C. and Borst P. (1972) The gel electrophoresis of DNA. *Biochim. Biophys, Acta* **269:** 192-200.

[2] Brunh, C.F. and Simpson, L. (1977) Comparison of various ultraviolet sources for fluorescent detection of ethidium bromide –DNA complexes in polyacrylamide gels. *Anal. Biochem.* **82:** 455-462.

[3] Helling R.B., Goodman H.M., and Boyer H.W. (1974) Analysis of endonuclease R-EcoRI fragments of DNA from lambdoid bacteriophages and other viruses by agarose-gel electrophoresis. *J. Virol.* **14:** 1235-1244.

[4] Holmes, D.L. and Stellwagen, N.C. (1990) The electric field dependence of DNA mobilities in agarose gels: A reinvestigation. *Electrophoresis* **11:** 5-15.

[5] Johnson P.H. and Grossman L.I. (1977) Electrophoresis of DNA in agarose gels. Optimizing separation of conformational isomers of double and single- stranded DNAs. *Biochemistry* **16:** 4217-4225.

[6] Krikpatrick F.H. (1990) Overview of agarose gel properties. *Curr. Commum. Cell. Mol. Biol.* **1:** 9-22.

[7] Lane, D., Prenti, P. and Chandler, M. (1992) Use of gel retardation to analyse protein nucleic acid interactions. *Microbiol. Rev.* **56:** 509-528.

[8] LePecq J.B. and Paoletti C. (1967) A fluorescent complex between ethidium bromide and nucleic acids: Physical-chemical characterization. *J. Mol. Biol.* **27:** 87-106.

[9] Lundstrom, R.C. and Roderick, S.A. (1979) Fish species identification by thin layer isoelectric focusing of sarcoplasmic proteins. *Science Tools* **26(3):** 38-43.

[10] Ornstein, L. (1964) *Ann. N.Y. Acad. Sci.* **121:** 321-349.

[11] Schwartz, D.C. and Cantor, C.R. (1984) Separation of yeast chromosome sized DNAs by pulsed field gradient gel electrophoresis. *Cell* **37:** 67-75.

[12] Sharp P.A., Sugden B., and Sambrook J. (1973) Detection of two restriction endonuclease activities in *Haemophilus parain-flucnza* using analytical agarose-ethidium bromide electrophoresis. *Biochemistry* **12:** 3055-3063.

[13] Southern E.M., Anand R., Brown W.R.A., and Fletcher D.S. (1987) A model for the separation of large DNA molecules by crossed field gel electrophoresis. *Nucleic Acids Res.* **15:** 5925-5943.

[14] Southern, E.M. (1975) Detection of specific sequences among DNA fragments separated by gel electrophoresis. *J. Mol. Biol.* **98:** 503-517.

[15] Southern, E.M. (1979) Gel electrophoresis of restriction fragments. *Methods Enzymol.* **68:** 152-176.

[16] Thorne, H.V. (1966) Electrophoretic separation of polyoma virus DNA from host cell DNA. *Virology* **29:** 234-239.

[17] Thorne, H.V. (1967) Electrophoretic characterization and fractionation of polyoma virus DNA. *J. Mol. Biol.* **24:** 203-211.

[18] Wieslander L. (1979) A simple method to recover intact high molecular weight RNA and DNA after electrophoresis separation in low gelling temperature agarose gels.*Anal. Biochem.* **98:** 305-309.

Chapter VI

Cloning Strategies

6.1. Introduction

DNA cloning strategies require four essential steps:

1. Method for generating DNA fragments
2. Reactions which join foreign DNA to the vector
3. A mean of introducing the artificial recombinant into a host cell in which it may replicate, and
4. A method of selecting or screening for a clone of recipient cells that has acquired the recombinant.

Suppose we have to clone a single copy gene from human genome, if human genomic DNA with a restriction endonuclease such as EcoRI, insert the DNA fragment into phage lambda vector and then attempt to isolate the desired clone. Human haploid genome is 2.8 x 10^6, assuming that EcoRI gives, on an average fragments of about 4 Kb, then we shall see that over 7 x 10^5 independent recombinants would be prepared and screened to have a reasonable chance of having the desired sequence. Thus we may have to obtain a very large number of recombinants which may contain a complete collection of all the DNA sequences in the entire human genome. Such a collection of clones from which we select the desired clone is called **Gene library** (**genomic DNA library**) or **gene bank**. Gene library is defined as the random collection of DNA fragments cloned in a vector that ideally includes all the genetic information of that species. Sometimes, it is also called **shot gun collection**.

6.2. Construction of Genomic DNA Libraries

The genomic DNA sequences produced by cutting the DNA with restriction endonucleases to generate the genomic DNA libraries may have some drawbacks:

a. The gene may cut one or more times internally by the restriction endonuclease (eg. EcoRI) with the result that it does not generate fragments of same size especially if the gene is too large in size.
b. It may be desired to obtain fragments from extensive regions flanking the gene or whole gene clusters.
c. Gene may be contained on an EcoRI fragment that is larger than the vector can accept, thus the desired gene may not be cloned at all.

These drawbacks may be overcome by cloning random DNA fragments of large size (20 Kb). Thus, clones produced from these random fragments may have DNA sequence over lapping one another giving an opportunity to walk from one clone to an adjacent clone. In addition, larger size of each cloned DNA will require fewer clones for preparation of complete genomic library.

Various methods are available for preparation of random DNA fragments like random breaking by mechanical shearing but more commonly used method involves restriction endonuclease partial digestion of genomic DNA. The target DNA is digested with a mixture of two restriction enzymes. These enzymes have tetranucleotide recognition sites, those occur frequently in the target DNA. A partial double-digest would produce fragments of average size less than 1 Kb. Therefore, to produce bulk fragments, preferably in the range of 10-30 Kb DNA only partial digestion of genomic DNA is carried out. These random set of overlapping DNA fragments can be fractionated either by velocity centrifugation on a sucrose gradient or by preparative gel electrophoresis so as to give a random population of about 20 Kb DNA fragments, which are suitable for insertion into the phage vector. Packaging *in vitro* ensures that an appropriately large number of independent recombinants can be recovered that will give genomic library representing complete DNA sequence. The use of two different restriction endonucleases with completely unrelated recognition sites, Hae III and Alu I, ensure fragmentation that is nearly random. Both of these enzymes produce blunt ends, and the cloning strategy requires linkers. A convenient alternate is to use a single restriction endonuclease which cuts frequently such as Sau 3A I. Though Sau 3A I creates a partial digest that is slightly less close to random than that achieved with two restriction enzymes mentioned earlier. However, it has the advantage that the Sau 3A I fragments can be readily inserted into high capacity phage vectors such as EMBL which has been digested with Bam H I because Sau 3A I and Bam H I create the same cohesive ends. These methods employing partial digestion of DNA coupled with packaging with the phage lambda recombinants have been widely used for creating genomic DNA libraries.

Cosmic vectors instead of the phage vectors may also be used. Though the cosmid vectors have high efficiency for packaging as well as even high capacity than any phage lambda vector, but there are two drawbacks with cosmid vectors. Most researchers find that screening libraries of phage recombinants by plaque hybridization gives better results than screening libraries of bacteria containing cosmid recombinants by colony hybridization. Plaques usually show lesser back ground hybridization than colonies. If an amplified genomic library is to be stored, with phage vector, the initial recombinant DNA population is packaged and plated out and it can be screened at this stage. The amplification is so great that sample of this amplified library could be plated out and screened with different probes on hundreds of occasions. The amplified library can be stored almost indefinitely as phages have a long shelf life. On the contrary, bacterial colonies containing cosmids and amplified library can be stored but there is a loss of viability on storage of such bacteria. Use of any amplified library may have possibility of distortion. Not all recombinants in a population will propagate equally well, eg. variation in target DNA size or sequence may affect replication of recombinant phage, plasmid or cosmid. Therefore, when a library is put through an amplification step, particular recombinant may reflect an increase in frequency or decrease in frequency, or total loss. With the help of modern vectors and cloning strategies, it is preferable to construct the new library for each screening rather than using a previously amplified library.

Although genomic DNA libraries in phage lambda vectors are expected to contain most of the sequences of the genome from which they have been derived but few reports suggest that some of the sequences that are present in eukaryotic genome are not found in genomic library. In addition, it has been observed that deletions can occur during cloning of mammalian DNA. Some of these deletions can be checked by growth on an *E. coli recA* recombination deficient host.

The inverted repeat sequences (large palindromes) can not be cloned in phage lambda vectors unless special host strains are used. It has been shown that a host mutant *rec*B, *rec*C and *sbc*B recombination-function genes will propagate many phage recombinants containing human DNA fragments that would not be propagated on other host strains.

6.3. Chromosomal Walking

Isolation of gene sequences whose function is quite unknown but whose genetic location is known can be achieved by the technique called Walking along the chromosome. This approach is based on:

1. A cloned genomic fragment must be found as a starting point for the walk. This should be as close as possible to the suspected destination point. In the human genome the starting point may be a restriction fragment length polymorphism (RFLP) sequence that is closely linked to a disease locus. In *Drosophila*, the starting point may be found cytologically by *in situ* hybridization to polytene chromosomes.
2. A random set of cloned genomic DNA is localized in this way and one is chosen whose location on the chromosome in question is closest to map position of the mutation under investigation.
3. The genomic library is then screened with this chosen clone as a probe to identify other clones containing DNA with which it reacts and that represent clones overlapping with it. The overlap can be to the left or to the right.
4. Repetition of this single walking step along the chromosome. The walk can potentially occur in both directions along the chromosome. Assembling contiguous clones can make a large scale map of the walk. This will be done in combination with methods for analyzing very large DNA regions, discussed below.

Repeated DNA sequences may occur dispersed at several places in the genome and could disrupt the orderly progress of the walk. Therefore, the probe may be used as overlapping from one genomic clone to the next but it must be a unique sequence clone, or a sub-clone which has been shown to contain only an unique sequence.

For the identification of the cystic fibrosis gene on chromosome 7 under human genome project, the following criteria were used to define whether a cloned DNA segment might contain the cystic fibrosis gene:

1. Identification of cross hybridization sequences from other species that relies on the tendency for gene sequences to be conserved in evolution, whereas 'spacer' DNA is not well conserved.
2. In vertebrate gene, 5' upstream region is often marked by CG rich dinucleotide hence identification of 'CpG islands' sequences rich in the CG dinucleotide was carried out.
3. Characterization of transcripts of the segment in normal tissues that are affected in the diseased state.
4. Isolation of cDNAs corresponding to the gene segment.
5. Identification of open reading frames in the nucleotide sequence of the segment.

Though, chromosomal walking appears too simple in principle, but technically walking on a large scale needs a lot of efforts. For large distances, it is usually combined with chromosome jumping. Highly repeated unclonable, or unstable genomic segments are characterized by chromosomal walking. Chromosomal walking has become a key technique for the human genome project as well as in studies of the *Drosophila* genome.

Following the advantages of chromosomal walking, improved vectors and strategies have been developed in response to the need for easier chromosome walking. For example, two phage derivatives, λDASH and λFIX have been designed by Stratagene Cloning Systems. It is a replacement vector and is capable to accommodate inserts of 9-22 Kb and includes multiple cloning sites. These sites define the stuffer fragment which bears red^+, gam^+ genes and therefore permits *Spi* selection. In this regard, λDASH resembles EMBL 3. T3 and T7 promoters for the T3 and T7 RNA polymerase are situated immediately adjacent to the cloning sites. These promoters allow RNA probes to be made from insert sequences. End specific RNA probes will be produced if the recombinant DNA to be transcribed into RNA is first digested with a restriction endonuclease that cuts more frequently (eg. a tetra-nucleotide target site), and thus leaves only a short insert DNA fragment attached to the T3 and T7 promoters. These probes are ideal for probing a library for the next step in a chromosome walk and have the great advantage that they can be made conveniently.

6.3.1. Long Distance Chromosome Walking and Analyzing Very Large DNA Sequences

Chromosome walks of thousands of kilobases correspond to a recombination distance of 1 cm (1% recombination) in human. Analyzing mapping and walking of such large distance has been overcome by three major developments:

1. The availability of restriction endonucleases as well as combinations of methylases and endonucleases which cut DNA at very frequent sites.
2. The development of pulsed field gel electrophoresis and field inversion gel electrophoresis to resolve very large DNA fragments.
3. Use of specialized chromosome jumping strategies.

6.3.2. Cutting DNA at Very Rare Target Sites

Not I (GCGGCCGC), Sfi I (GGCCNNNNNGGCC) and Pac I (TTAATTAA) are type II restriction endonucleases that cut octanucleotide recognition sites and these sequences are expected to cut DNA very infrequently. In addition, there are restriction endonucleases with shorter recognition sequences that have the property of being uncommon sequences in mammalian DNA. Examples of this category are Nru I (TCGCGA) and Bss H II (GCGCGC). Their target sequences include two CG dinucleotides. The CG dinucleotides are rare in mammalian DNA (except at CpG islands) and the target sites for these enzymes are consequently very infrequent.

In target sites additional specificities may be created by combining the specificity of certain methylases with that of restriction endonucleases. An example of this approach involves the restriction endonuclease, Dpn I, which cuts DNA at the sequence G A^m TC to produce flush ends. The enzyme is unusual as it requires the methylation of the adenines in both strands of the DNA for cleavage to occur. Thus DNA from *dam*$^+$ *E. coli* is cleaved by Dpn I, whereas DNA from *dam*$^-$ mutants is not. This property has been exploited by combining the specificity of the modification methylases M. Taq I or M. Cla I. The M. Taq I methylates both the strands of the sequence TCGA to create $TCGA^m$. Thus in DNA that is not previously methylated at these sites, cleavage by Dpn I will only occur at the following octanucleotide sequence: $TCGA^m$ TCGA. Similarly, M. Cla I methylates both strands of the sequence ATCGAT to produce ATCGAm T, and so in combination with Dpn I cleavage will only occur at the decanucleotide sequence, ATCGATCGAT.

6.3.3. Pulsed Field Gel Electrophoresis and Field Inversion Gel Electrophoresis (FIGE)

A new type of gel electrophoresis is in use which resolves DNA molecules up to 2000 Kb in length. The technique uses conventional agarose and conventional buffers. The innovation is the use of alternately pulsed, perpendicularly (orthogonally) oriented electrical fields. The separation appears to depend upon the electrical perturbation of the orientation of the DNA, and on the degree of extension of long DNA molecules. The refixation time of such molecules in free solution is a very sensitive function of the molecular weight. This orthogonal pulsed field gel electrophoresis (PFGE) technique has been further developed, but has the disadvantages that the DNA samples do not run in straight line tracks.

Field inversion gel electrophoresis is a technique that overcomes these difficulties. This resolves DNA of large fragments up to 2000 Kb without the need for a complicated perpendicular field. It uses a conventional electrophoresis apparatus with an electric field that pulses forward reverse combined with a pause between each phase. The pulses are in the form of a time ramp. The DNA tracks run in straight lines and are comparable across the gel, thus making calibration with markers and the interpretation of Southern blots much simpler as compared to orthogonal PFGE.

6.3.4. Chromosome Jumping Strategies

Chromosomal jumping depends upon the circularization of very large genomic DNA fragments followed by cloning DNA from the region covering the closure site of these circles, thus bringing together DNA sequences that were originally located at a considerable distance apart in the genome. These cloned DNAs from the closure sites make up a jumping library. Such a jumping library greatly speeds up the process of long range chromosomal walking.

A strategy for creating a jumping library using Not I digested human genomic DNA has been suggested. This enzyme cuts only rarely in mammalian DNA, and so drastically reduces the size of the library required to cover the mammalian genome. This strategy has the additional advantage that analysis of jumps is easy by FIGE analysis and Southern blotting of Not I digested genomic DNA. However, with a complete digest such as this overlaps between Not I fragments have to be obtained indirectly. One solution is to use 'linking clones' containing conventionally cloned fragments with an internal Not I site. This technique has been applied to a jump of 100 Kb in the cystic fibrosis locus and to a jump of 200 Kb in the region of the Huntington disease gene.

6.3. cDNA Library / cDNA Cloning

The quality of cDNA library depends upon the quality of the mRNA. Pure, undegraded mRNA is essential for the construction of large, representative cDNA libraries. Secondary structure of mRNA molecules can cause the synthesis of truncated cDNA fragments. In this case, treatment of the mRNA with a denaturant such as methyl mercuric hydroxide, prior to synthesis may be necessary.

Cloned eukaryotic cDNAs have advantage of lacking the intron sequence that are usually present in the corresponding genomic DNA. Introns are non-coding sequences that often occur within eukaryotic gene sequences. They can be situated within the coding sequence itself, where they interrupt the co-linear relationship of the gene with the encoded polypeptide. They may also occur in the 5' or 3' untranslated regions of the gene, but in any event they are copied into RNA by RNA polymerase when it transcribes the gene. The initial primary transcript is a precursor to mRNA. It goes through a series of processing events in the nucleus before appearing in the cytoplasm as mature mRNA. These events include the removal of intron sequences by a process called *splicing*. When cDNA is derived from it, lacks intron sequences. Since removal of eukaryotic intron transcripts by splicing does not occur in bacteria, eukaryotic cDNA clones find application where bacterial expression of the foreign DNA is necessary, either as prerequisites for detecting the clone, or because the polypeptide product is the primary objective. In situations where the sequence of the genomic DNA is known, the position of intron/exon boundaries can be assigned by comparison with the cDNA sequence.

cDNA cloning is carried out for the analysis of temporally regulated gene expression in development, or tissue specific gene expression. Using the differential screening procedure it is possible to screen a cDNA clone library to identify cDNA clones derived from mRNA molecules present in one cell type but absent in another cell type. Another strategy for harvesting abundant cDNA involves the cloning of the cDNA directly in an M13 vector such as M13mp8. A set of clones can then be sequenced and identified on the basis of the polypeptide encoded by clones. The DNA sequences of 178 randomly chosen muscle cDNA recombinants has been determined by a strategy called as shot gun sequencing. Complete amino acid sequences has also been determined for 19 abundant muscle related proteins.

For cDNA clones in the low abundance class it is usual to construct a cDNA library. By packaging *in vitro* phage lambda vectors as these vectors yield large numbers of cDNA clones. Insertion vectors such as lambda gt10, lambda NM1149, lambda zap or lambda gt11 are particularly well suited for such cDNA cloning.

A selective strategy for cDNA cloning involves specific priming of cDNA synthesis with oligonucleotide primers (either unique or degenerate in sequence). Selective priming is also used in various PCR strategies that involve standard PCR protocols. In general, the most commonly used primary screening technique involves colony or plaque hybridization with radioactive or immunochemical probes. This is applicable to very large libraries and the effort involved is largely independent of the number of recombinants to be screened. Therefore, enrichment of the starting mRNA in order to reduce the number of recombinants to be screened has little advantage.

6.4.1. Full Length cDNA Cloning

In order to make a full length cDNA library, cDNAs which contain both ends of the intact mRNA should be selected. For that purpose, features which are characteristic of the 3' end and the 5' end of the mRNA should be used as 'tag'. The full length cDNA can then be selected through selection steps for both the 3' end and the 5' end 'tags'. The poly A stretch is a characteristic feature of the 3' end of mRNA. Conventional methods use the Poly A as a sequence tag to select the 3' end of mRNA. In the conventional methods, the first strand cDNA is usually synthesized using an oligo dT primer. The oligo dT primer hybridizes mostly at the poly A, most of the cDNAs are selectively synthesized from the 3' end of the mRNA. Thus, the conventional methods include a selection step for the 3' end of mRNA. As a result, the largest part of the cDNA library consists of cDNAs which lack the 5' end of mRNA. The main reason for this lies in the fact that mRNA does not originally have a sequence tag at the 5' end of mRNA but has a characteristic structure called cap structure which is not a sequence tag. If the 5' end of the mRNA were also a sequence tag, it would be easier to use it to select the 5' end of mRNA.

The full length cDNA which contains the entire sequence of the mRNA, is the ultimate goal of the cDNA cloning. Unfortunately, cDNA libraries constructed by many types of conventional methods have a high content of non-full length cDNA clones. One of the reasons for this high content is that reverse transcriptase tends to stop during the first strand synthesis and to fall off leaving non-full length cDNA. Thus, non-full length cDNA is an inevitable result of the use of reverse transcriptase for the cDNA synthesis.

Early strategies for synthesizing cDNA molecules for cloning involved self priming of second strand synthesis and therefore led to the loss of some sequence corresponding to the 5' end of the mRNA. Several strategies were developed to overcome this difficulty. A method has been developed in which dC tailing of single stranded cDNA followed by Oligo dG priming of second strand synthesis did not lead to hairpin formation, nuclease S1 treatment also did not require, and consequently this is an effective method for generating full length cDNA clones.

In another method, use of nuclease S1 has been eliminated and the oligo dT sequence for priming the first strand cDNA synthesis is linked to the vector DNA in a prior reaction. The method has been reported to promote full length cDNA cloning with a very high efficiency. It employs a second strand synthesis step in which the RNA strand is replaced in a DNA polymerase reaction that is primed by nicking the RNA with RNase H. The second strand synthesis occurs at these nicks by a nick translation type of reaction. This type of second strand reaction has been exploited by other researchers in efficient cDNA library cloning schemes. A particular advantage of the strategy is that the cDNA is inserted into the vector in a defined orientation. This is useful in those strategies in which the cDNA is inserted into a vector adjacent to phage T3, T7 or SP6 promoter because then an RNA copy can be synthesized *in vitro* from a defined cDNA strand with purified phage RNA polymerase. With present technique of cDNA cloning, the yield of recombinant DNAs that have full length DNA sequences that is cDNA containing the entire nucleotide sequence of the mature mRNA is low. Generally, most cDNA clones contain 3' untranslated and variable amounts of the protein coding sequence, therefore to obtain cloned cDNAs with complete 5' untranslated and protein coding sequences is difficult, particularly if the mRNA codes for a large protein. Although such truncated cDNAs are still useful as hybridization probes, but they can not direct the synthesis of complete proteins after their introduction into bacterial or mammalian cells. The cDNA cloning methods have been developed that increases the probability of obtaining the recombinants with full length cDNA inserts.

In earlier methods, the initial cDNA copy of mRNA was being synthesized with reverse transcriptase using a primer oligo dT annealed to poly A tail. The quality of reverse transcriptase, the integrity and the secondary structure of the mRNA as well as reaction conditions influence the length of the primary reverse transcript and therefore the completeness of the subsequently cloned cDNA.

The modification in cDNA cloning procedure that permits the recovery in high yields of plasmid recombinants with full length cDNA is described here. This method uses a plasmid DNA vector which itself serves as a primer for the first and second strand cDNA synthesis and one of the steps is designed to enrich recombinants containing full length cDNA over those with truncated DNAs. The procedure is applied for the cloning of α and β globulin cDNA from rabbit reticulocyte mRNA.

The method is also applicable to purified or mixed mRNA, to poly (A) + viral RNAs or to cellular RNAs to which poly A tails can be added at the 3' end. The procedure is relatively rapid taking less than two days from reverse transcription to the collection of transformants containing the cloned cDNAs. Moreover, it uses commercially available reagents and enzymes and standard biochemical procedures. Aside from the relative ease and rapidity, relative to those previously used, is the high efficiency of cloning full or nearly full length cDNAs. Applying the procedure to rabbit reticulocyte mRNA yielded approximately 10 globulin cDNA clones per microgram of mRNA, of which 10% had the complete mRNA sequence at the 5' end and 20% to 40% had all but two or three short at the 5' most nucleotides of the globulin mRNA sequences. As anticipated, incorporation of one of the complete α-globulin cDNA segments into the vector SVTG5 yielded SV40 like late mRNAs containing the α-globulin cDNA sequence and immunoprecipitatable α-globulin protein. The factors contributing to the relatively high cloning efficiency of full length cDNAs are listed below:

1. Priming by the plasmid DNA for the synthesis of the first cDNA to the vector.
2. Selective cloning of complete or nearly complete cDNA copies of the mRNAs is necessary because the oligo dC tails need to effect ring closure and complete second strand cDNA are synthesized more efficiently by terminal transferase with substrates containing base paired, full length reverse transcripts than with shortened or truncated cDNA strands base paired to longer mRNAs.
3. Replacement of the mRNA by a second strand is effected by the actions of *E. coli Pol* I-mediated nick translation (to replace RNA with DNA sequence), and *E. coli* DNA ligase (to join the DNA fragments into the continuous second cDNA strand).

Currently, nucleic acid hybridization or functional assays for the mRNA are the sole means to identify specific cDNA clones. By appropriate choice of the vector primer and linker DNAs, possibility of the construction of recombinants that can express the cloned cDNAs directly in either bacterial or mammalian cells is being explored.

6.4.2. PCR Technology for Generating cDNA Clones

In this chapter, methods have been described for constructing genomic and cDNA libraries. However, there are alternative routes to obtain specific genomic or cDNA clones which are based upon the polymerase chain reaction (PCR). For many applications, a PCR based approach is quicker and simpler than library's construction and screening. Even in circumstances where a suitable library already exists, a PCR-based approach may be attractive.

6.4.3. The Use of Random Primers in cDNA Cloning

In the example discussed so far, cDNA synthesis has been initiated on poly (A)+ mRNA by priming with oligo dT sequences but this method has three limitations:

- Not all RNAs bear a 3' terminal poly A sequence. It is possible to add a poly A sequence *in vitro* with purified poly A polymerase but this may be problematic.
- Large mRNAs can be difficult to deal with because it is not reasonable to expect to synthesize and clone sequence which lie further than a few kilobase from the oligo dT primer.
- Because priming occurs at the 3' end of poly A + mRNAs and cDNA synthesis is often incomplete, cDNA libraries are enriched in 3' terminal sequences.

The first two limitations are common in cloning genomic RNAs from RNA viruses. The third limitation may be important where cDNA libraries are to be made in a vector such as λ gt 11 or zap and screened for expression of fusion polypeptides. These limitations can often be overcome by priming the first strand cDNA synthesis with random oligonucleotide primers and not with oligo dT. Usually the primer consists of a mixture of all possible chemically synthesized hexadeoxynucleotides. These hybridize at random sites along the RNA, prime cDNA synthesis, and generate cDNA sequences of sufficient sizes to be useful when cloned.

References

[1] Carle, G.R. and Olson, M.V. (1984) Separation of chromosomal DNA molecules from yeast by orthogonal-field alteration gel electrophoresis. *Nucleic Acid Res.* **12:** 5647-5664.

[2] Collins, F.S. and Weissman, S.M. (1984) Directional cloning of DNA fragments at a large distance from an initial probe: A circularization method. *Proc. Natl. Acad. Sci. USA.* **81:** 6812-6816.

[3] Gubler, U. and Hoffman, B.J. (1983) A simple and very efficient method for generation cDNA libraries. *Gene* **26:** 263-269.

[4] Hanahan, D. and Meselson, M. (1980) Plasmid screening at high colony density. *Gene* **10:** 63-67.

[5] Heidecker, G. and Messing, J. (1983) Sequence analysis of zein cDNAs by an efficient mRNA cloning method. *Nucleic Acid Res.* **11:** 4891-4906.

[6] Land, H., Grez, H., Hauser, H., Lindenmaier, W. and Schutz, W. (1981) 5'-Terminal sequences of eukaryotic mRNA can be cloned with high efficiency. *Nucleic Acid Res.* **9:** 2251-2265.

[7] Lapeyre, B. and Amatric, F. (1985) A powerful method for the preparation of cDNA libraries: Isolation of cDNA encoding a 100 Kda nuclear protein. *Gene* **37:** 215-220.

[8] Leach, D.R.F. and Stahl, F. (1983) Viability of lambda phages carrying a perfect palindrome in the absence of recombination nucleases. *Nature* **305:** 448-451.

[9] Maniatis, T., Hardison, R.C., Lacy, E., Lauer, J., O'Connell, C., Quon, D., Sim, G.K. and Efstratiadis, A. (1978) The isolation of structural genes from libraries of eukaryotic DNA. *Cell* **15:** 687-701.

[10] Okayama, H. and Berg, P. (1982) High efficiency cloning of full length cDNA. *Mol. Cell Biol.* **2:** 161-170.

[11] Poustka, A. and Lehrach, H. (1986) Jumping libraries and linking libraries: The next generation of molecular tools in mammalian genetics. *Trends in Genetics* **2:** 174-179.

[12] Putney, S.D., Herlihy, W.C. and Schimmel, P. (1983) A new troponin T and cDNA clones for 13 different muscle proteins found by shot gun sequencing. *Nature* **302:** 718-721.

[13] Richards, J.E., Gilliam, T.C., Cole, J.L., Drumm, M.L., Wasmuth, J.J., Gusella, J.F. and Collins, F.S. (1988) Chromosome jumping from D4S10 (G8) toward the Huntington disease gene. *Proc. Natl. Acad. Sci. USA.* **85:** 6437-6441.

[14] Rommens, J.M., Iannuzzi, M.C., Kerem, B., Drumm, M.L., Melmer, G., Dean, M., Rozmahel. R., Cole, J.L., Kennedy, D., Hidaka, N., Zsiga, M., Buchwald, M., Riordan, J.R., Tsui, L-C and Collins FS (1989) Identification of the cystic fibrosis gene: Chromosome walking and jumping. *Science* **245:** 1059-1065.

[15] Schwartz, D.C. and Cantor, G.R. (1984) Separation of yeast chromosomal sized DNAs by pulse field gradient electrophoresis. *Cell* **37:** 67-75.

[16] Wyman, A.R., Wolfe, L.B. and Botstein, D. (1985) Propagation of some human DNA sequences in bacteriophage λ vectors requires mutant *Escherichia coli* hosts. *Proc. Natl. Acad. Sci. USA.* **82:** 2880-2884.

Chapter VII

DNA Sequencing Techniques

7.1. Introduction

DNA sequencing is the determination of the precise sequence of nucleotides in a sample of DNA. The DNA sequencing methods have revolutionized our understanding of gene structure, and it is now routine to sequence any newly isolated DNA fragment of interest. In molecular cloning laboratories, DNA sequencing today is used chiefly to characterize newly cloned cDNAs, to confirm the identity of a clone or mutation to identify polymorphism and mutations in genes of particular interest. Until few years ago, these tasks were achieved by manual laborious sequencing protocols. However, nowadays, most laboratories have access to sequencing facilities equipped with oligonucleotide synthesizers and automated machines that utilize fluorescent detection systems rather than autoradiography to visualize the products of DNA sequencing reactions. The sequences are documented automatically and transferred to computers.

In the mid 1970s, when molecular cloning techniques in general were rapidly improving, simple methods were developed to determine the nucleotide sequence of DNA. The first attempts were based on techniques developed by Robert Holley's group to sequence tRNAs. Holley's method involved digesting RNA with sequence specific RNAses, fractionating the resulting oligonucleotides by ion-exchange chromatography or two dimentional chromatography, and establishing the order of bases within each fragment by exonuclease digestion. By using two different methods of fragmentation, Holley was able to establish overlaps between fragments and hence to assemble the entire 77-nucleotide sequence of a yeast alanine tRNA.

Nowadays, due to rapid developments in molecular biology, DNA sequencing has become easier due to following advancements:

- The development of restriction endonucleases
- The development of gel electrophoretic techniques, which can separate DNA fragments differing by one nucleotide only
- The latest gene cloning and PCR methods which can make large DNA fragments
- The development of two easy and rapid DNA sequencing procedures.

The best known DNA sequencing techniques are the enzymatic method of Sanger developed at Cambridge, England and the chemical degradation method of Maxam and Gilbert worked out at Gilbert's laboratory, Harvard University, USA. Although very different in principle, both these two methods generate populations of oligonucleotides that begin from a fixed point and terminate at a particular type of residue. In the simplest case, four populations are created that terminate at A, G, C, and T residues, respectively. The termination points are nucleotide specific but occur randomly along the length of the target DNA. Each of the four populations therefore consists of a nested set of fragments whose lengths are determined by the distribution of a particular base along the length of the original DNA. These populations of fragments are resolved by electrophoresis under conditions that discriminate between individual DNAs differing in length by as little as one nucleotide. When the four populations are loaded into adjacent lanes of a sequencing gel, the order of nucleotides along the DNA can be read directly from the image of the gel.

Fredrick Sanger was first accredited with the invention of DNA sequencing techniques. During experiments to uncover the amino acids in the insulin protein, he developed the basics of modern sequencing methods. Sanger's approach involved copying DNA strands with dNTPs (deoxy ribonucleoside triphosphates) and terminating them with tagged ddNTPs (dideoxyribonucleoside triphosphates) which would show the location of nucleotides in the strands through autoradiography technique. Almost simultaneously, Alan Maxam and Walter Gilbert were creating a somewhat different method that relies on the base specific chemistry. Later, there have been several modifications of chain termination method and automated sequencing technique has been developed. This chapter describes various methods used for DNA sequencing.

7.2. DNA Sequencing by Maxam and Gilbert Method

DNA can be sequenced by a chemical procedure that breaks the terminally labeled DNA molecule partially at each repetition of a base. The lengths of the labeled fragments then identify the positions of that base. Base specific reactions that cleave DNA are carried out preferentially at guanines, at adenines, at cytosines and thymines, and at cytosines alone. In this Maxam and Gilbert method, the starting point is a defined labeled end of a DNA restriction fragment. This method for DNA sequence makes use of chemical reagents to bring about base specific cleavage of the DNA. The DNA strands to be sequenced must be radioactively labeled at one end with a ^{32}P-phosphate group.

Reagents for Maxam and Gilbert Sequencing

When the products of these four reactions are resolved by size using electrophoresis on a polyacrylamide gel, the DNA sequence can be read from the pattern of the radioactive bands. The technique permits sequencing of at least 100 bases from the point of labeling.

Base specificity	Base reaction	Altered base removal	Strand cleavage
G	Dimethyl sulfate	Piperidine	Piperidine
G + A	Acid	Acid catalyzed Depurination	Piperidine
T + C	Hydrazine	Piperidine	Piperidine
C	Hydrazine + NaCl	Piperidine	Piperidine
A>C	NaOH	Piperidine	Piperidine

DNA (a defined restriction fragment) is attached with reagents that first damage and then remove base from its sugar. The exposed sugar easily breaks and an alkali or amine catalyzed series of β-elimination reactions will cleave the sugar completely from its 3' and 5' phosphates. The reaction with the bases is a limited one, damaging only one residue for every 50 to 100 bases along the DNA. The second reaction to cleave the DNA strand must go to completion, so that the molecules finally analyzed do not have hidden damages. The purine specific reagent is dimethyl sulfate and the pyrimidine specific reagent is hydrazine.

The sequencing requires DNA molecules, either double stranded or single stranded that are labeled at one end of one strand with ^{32}P. This can be a 5' or 3' label. A restriction fragment of any length is labeled at both ends for example, by being first treated with alkaline phosphatase to remove terminal phosphates and then labeled with ^{32}P by transfer from γ-labeled ATP with polynucleotide kinase. There are two strategies:

1. The double stranded molecule is cut by a second restriction endonuclease and the two ends resolved on a polyacrylamide gel and isolated for sequencing.
2. The double labeled molecule is denatured and the strands are separated on a gel, extracted, and sequenced.

The Chemical Basis

Base Specific Cleavage

Guanine Cleavage

Dimethyl sulfate methylates the guanines in DNA at the N7 position and the adenines at the N3. The glycosidic bond of a methylated purine is unstable and breaks easily on heating at neutral pH, leaving the sugar free. Treatment with 0.1 M alkali at 90°C will leave the sugar from the neighboring phosphate groups. When the resulting end labeled fragments are resolved on a polyacrylamide gel, the autoradiograph contains a pattern of dark and light bands. The dark bands arise from breakage at guanines, which are methylated 5 fold faster than adenines. This strong guanine/weak adenine pattern contains almost half the information necessary for sequencing. However ambiguities can arise in the interpretation of this pattern because the intensity of isolated bands is not easy to assess. To determine the bases, we compare the information contained in this column of the gel with that in a parrallel column in which the breakage at the guanines is suppressed, leaving the adenines apparently.

Adenine Cleavage

The glycosidic bond of methylated adenine is less stable than that of methylated guanosine, thus, gentle treatment with dilute acid releases adenines preferentially. Subsequently cleavage with alkali then produces a pattern of dark bands corresponding to adenines with light bands at guanines.

Cleavage at Cytosine and Thymine

Hydrazine acts with thymine and cytosine, cleaving the base and leaving ribosylurea. Hydrazine then may react further to produce a hydrazone. After a partial hydrazinolysis in 15 to 18 M aqueous hydrazine at 20°C, the DNA is cleaved with 0.5 M piperidine. This cyclic secondary amine, as the free base, displaces all the products of the phosphate. The final pattern contains bands of similar intensity from the cleavages at cytosine and thymine.

Cleavage at Cytosine Alone

The presence of 2M sodium chloride preferentially suppresses the reaction of thymine with hydrazine. Then, the piperidine breakage produces band only from cytosine.

Autoradiography

Freeze the gel for autoradiography. Remove one glass plate, wrap the gel and supporting plate with saran wrap, and mark the positions of the dyes with ^{14}C containing ink. Place the gel in contact with film in a light tight X-ray exposure holder (backed with lead and aluminum) at –20°C under pressure from lead bricks. The sequence can easily be read from the pattern of bands. The spacing between fragments decreases from the bottom toward the top of the gel. The slight variations in the spacing are sequence specific and reflect the last nucleotide added, a T or G decreasing the mobility more than an A or C. The fragments on the gel end with the base just before the one destroyed by the chemical attack.

The chemical sequencing method has certain specific advantages. First, the chemical treatment is easy to control. The chemical attack hits one base per strand, produces a rather even distribution of labeled material across the sequence. Second, each base is attacked, so that in a turn of any single base all those are displayed. The chemical distinction between the different bases is clear, and the sequence of both strands provides a more than adequate check.

7.3. The Sanger's Method of DNA Sequencing by Dideoxy Mediated

Chain Determination

General Principle

A synthetic oligonucleotide primer is annealed to a single stranded DNA template. Four different sequencing reactions are set up each containing a DNA polymerase and the four normal dNTPs. One of the precursors or, in some cases, the primer is labeled radioactively with ^{32}P, ^{33}P, ^{35}P or ^{35}S or with a nonradioactive fluorescent tag. The four reactions also contain a small proportion of a 2',3'-ddNTP that carries a 3'-H atom on the deoxyribosyl moiety. If the ddNTP molecule is incorporated into a growing DNA chain, the absence of a 3'-OH group prevents formation of a phosphodiester bond with the succeeding dNTPs. Further extension of the growing chain is impossible. Thus, when a small amount of one of the ddNTPs is included with the four conventional dNTPs in a reaction mixture for DNA synthesis, there is competition between extension of the chain and infrequent, but base specific, termination. The products of the reaction are a population of oligonucleotide chains whose lengths are determined by the distance between the 5' terminus of the primer used to initiate DNA synthesis and the sites of chain termination. In a sequencing reaction containing DNA, for example, the termination points correspond to all positions normally occupied by a deoxyadenosyl residue. By using the four different ddNTPs in four separate enzymatic reactions, populations of oligonucleotides are generated that terminate at positions occupied by every A, C, G, or T in the template strand. These populations of oligonucleotides can be separated by electrophoresis, and the locations of each band can be ascertained by autoradiography of the sequencing gel. The sequence of the newly synthesized strand can be read in a 5',3' orientation by calling the order of bands from the bottom to the top of the gel.

7.4. The Dideoxy Method of DNA Sequencing

The Sanger's technique uses controlled synthesis of DNA to generate fragments that terminate at specific points along the target sequence. The dideoxy method gets its name from the critical role played by synthetic nucleotides that lack the OH at the 3' carbon atom. A dideoxynucleotide can be added to the growing DNA strand. Strand elongation stops because there is no 3' OH for the next nucleotide to be attached to. For this reason, the dideoxy method is also called the **chain termination method**.

This method is similar to the 'plus and minus' method but makes use of the 2',3'-dideoxynucleoside triphosphates, which act as specific chain terminators of DNA and this method is more rapid and accurate than either the plus or the minus method.

The 'plus and minus' method is a relatively rapid and simple technique that has made possible the determination of the sequence of the DNA. It depends on the use of DNA polymerase to transcribe specific regions of the DNA under controlled conditions. Although the method is considerably more rapid and simple than other available techniques, neither the 'plus' nor the 'minus' method is completely accurate and in order to establish a sequence both must be together and sometimes confirmatory data are necessary. Recently, an another method has been developed which involves ribo-substitution. This method has certain advantages over the plus and minus method but this has not yet been extensively exploited.

Principle of the Method

It has been shown that the inhibitory activity of 2',3' dideoxythymidine triphosphate (ddTTP) on DNA polymerase I depends on its being incorporated into the growing oligonucleotide chain in place of thymidylic acid (dT). Because the ddT contains no 3'-hydroxyl group, the chain can not be extended further, so that termination occurs specifically at positions where dT should be incorporated. If a primer and template are incubated with DNA polymerase in the presence of a mixture of ddTTP and dTTP, as well as the other three deoxyribonucleoside triphosphates (one of which is labeled with ^{32}P), a mixture of fragments all having the same 5' and with ddT residues at the 3' ends is obtained. When this mixture is fractioned by electrophoresis on denaturing polyacrylamide gels, the pattern of bands shows the distribution of dTs in the newly synthesized DNA. By using analogous terminators for the other nucleotides in separate incubations and running the samples in parallel on the gel, a pattern of bands is obtained from which the sequence can be read off as in other rapid techniques mentioned above.

Two types of terminating nucleoside triphosphates have been used – the dideoxy derivatives and the arabinonucleosides. Arabinose is a stereoisomer of ribose in which the 2'-hydroxyl group is in different orientation. The arabinosyl nucleotides act as chain terminating inhibitors of *E. coli* DNA polymerase I in a manner comparable to ddT. However, synthesized chains ending in 3' arabinosyl nucleotide can be further extended by some mammalian DNA polymerases. In order to obtain suitable pattern of bands of which an extensive sequence can be read, it is necessary to have a ratio of terminating nucleoside triphosphate to normal nucleoside triphosphate such that only partial incorporation of the terminator occurs. For the dideoxy derivatives the ratio is about 100, and for the arabinosyl derivatives about 5000.

Analysis

The method described here has a number of advantages over plus and minus methods. First, it is simpler to perform because it requires no preliminary extension, thus avoiding incubation and purification on a Sephadex column. It requires only the commercially available DNA polymerase I (Klenow fragment). The results appear to be more clear cut with fewer artifact bands, and can usually be read further than with the plus and minus methods. Intermediate nucleotides in 'runs' show up as bands, thus avoiding source of error in the plus and minus method-estimating the number of nucleotides in a run. Theoretically one would expect the different bands in a run to be of the same strength, but this is not always the case. Frequently, the first nucleotide is the strongest. But in the case of ddGTP the second is the strongest. The reason for these effects is not understood, but they do not usually cause difficulties with deducing the sequences. For the longer sequences in which the separate bands in a run are not resolved, experience has shown that it is frequently possible to estimate the number of nucleotides from the strength and width of the band.

In the course of *in vitro* experiments to prepare DNA copies of high specific radioactivity, Sanger and his colleagues noticed that the size of newly synthesized DNA chains was frequently less than unit length. Premature termination generally occurred at the residue immediately upstream of the site at which a radioactive nucleotide should have been incorporated. This anomaly occurred because the concentration of the radioactive labeled nucleotide in the reaction was far lower than that of the other three unlabeled nucleotides. The 3' ends of the resulting mixture of products therefore corresponded to the position of the nucleotide whose concentration was limiting. By carrying out four separate reactions in which each of the nucleotides was limiting in turn and separating the four sets of truncated fragments according to their size, Sanger and his colleagues were able to generate a readable DNA sequence. Although the plus and minus technique was slow and inaccurate, it was used successfully to determine almost all of the 5386 nucleotide sequence of bacteriophage ϕ X 174 DNA. The plus and minus system was an important advancement because it first described:

1. The use of a specific primer for extension by DNA polymerase
2. Base specific chain termination, and
3. The use of polyacrylamide gels to discriminate between single stranded DNA chains differing in length by a single nucleotide.

The breakthrough came with the introduction of dideoxynucleoside triphosphates (ddNTPs)- nucleotide analogs that:

1. Could be substituted for deoxynucleotides at random positions during template – directed copying by a DNA polymerase and
2. Could efficiently terminate DNA synthesis in a base specific manner

By using ddNTPs corresponding to the conventional nucleotides in separate synthetic reactions, four populations of oligonucleotides could be generated that terminate at every position in the template strand. The DNA bands in the reactions could be displayed by gel electrophoresis and the DNA sequence deduced from the order of bands in the four lanes, these chain terminators worked so well that by the beginning of the 1980s, the Sanger method became the method of choice for DNA sequencing-a position it still occupies.

7.5. Sequencing by the Chain-Terminator or Dideoxy Procedure

This method is now favored for large scale DNA sequence determination. It was developed in conjunction with cloning the DNA to be sequenced in M13mp series of single stranded vectors (or alternately pEMBL vector may also be used).

In order to appreciate the elegance of the combined technology, it is necessary first to understand the chain terminator DNA sequencing procedure. This procedure capitalizes on two properties of DNA polymerase. First, its ability to synthesize faithfully a complementary copy of a single stranded DNA template. Second, its ability to use 2',3' dideoxynucleoside triphosphates as substrates. Once the analogue is incorporated at the growing point of the DNA chain, the 3' end lacks a hydroxyl group and no longer is substrate for chain elongation, thus the growing chain is terminated.

In practice, the Klenow fragment of DNA polymerase I which lacks the 5',3' exonuclease activity of the intact enzyme is used to synthesize a complementary copy of the single stranded target sequence. Initiation of DNA synthesis requires a primer and usually this is a chemically synthesized oligonucleotide which is annealed close by. DNA synthesis is carried out in the presence of the four deoxynucleoside triphosphates, one or more of which is labeled with ^{32}P, and in four separate incubation mixes containing a low concentration of one each of the four dideoxynucleoside triphosphate analog. Therefore, in each reaction there is a population of partially synthesized radioactive DNA molecules, each having a common 5' end, but each varying in length to a base specific 3' end. After a suitable incubation period, the DNA in each mixture is denatured, electrophoresed side by side, and the radioactive bands of single stranded DNA detected by autoradiography. The sequence can then be read off directly from the autoradiograph.

The DNA to be sequenced is first cloned into one of the clustered cloning sites in the *lac* region of M13mp series of vectors. Recombinants are detected by the formation of white plaques on media containing IPTG and X-gal. Virus is isolated from these white plaques, stock prepared and the single stranded viral DNA extracted for sequencing. A feature of M13mp series is that cloning into the same, specific region of the genome obviates the need for many different primers, since a single primer can be used for all inserts. The original primer was a short restriction of the fragment which was cloned in pBR322 and which was complementary to a region of the *lac* Z gene immediately adjacent to the right hand EcoRI insertion site. A suitable 14 base synthetic oligonucleotide was developed which was present just to the right of the polylinker. This was subsequently found to have low homology with a second site in M13, so that an improved 17 base oligonucleotide primer has been synthesized. This has no homology at any secondary site.

7.6. Modifications of Chain Terminator Sequencing

The combination of chain terminator sequencing, and M13 vectors to produce single stranded DNA, is powerful. Excellent quality sequence data is obtainable with this technique, especially when the improvements given by ^{35}S-labeled precursors and T7 DNA polymerase are exploited. Further modifications allow sequencing of double stranded DNA i.e. double stranded input DNA is denatured by alkali, neutralized, and one strand is then annealed with a specific primer for the actual chain terminator sequencing reaction. This approach has gained in popularity, as the convenience of having a universal primer has grown less important with the wide spread availability of oligonucleotide synthesizers. With this development, the chain terminator technique was liberated from its attachment to the M13 cloning system; for example, PCR amplified DNA segment can be sequenced directly, one variant of the double stranded approach, often employed in automated sequencing, is cycle-sequencing. This involves a linear amplification (compared to PCR) of the sequencing reaction using 25 cycles of denaturation, annealing of a specific primer to one strand only, and extension in the presence of Taq DNA polymerase plus labeled (radioactive or fluorescent) dideoxynucleotides.

Some scientists have recommended the use of 7-deaza-2'-deoxyadenosine-5'-triphosphate (C^7dATP) in place of dATP. In sequencing gel, it is expected that two DNA fragments having a difference of one nucleotide will separate. However, in certain circumstances, fragments differing in size by one or a few nucleotides migrate with very similar mobility, resulting in localized regions of reduced band spacing and parallel bands. This gel associated phenomenon has been called as **'band compression'**. Band compressions are independent of the DNA polymerase and reaction conditions used in DNA sequencing reactions. In fact, band compressions occur due to the presence of secondary structures in DNA that are not fully denatured during electrophoresis. Fragments that retain some secondary structure display increased mobility and appear as shorter in length. The secondary structures may involve in intra-strand hairpin structures formed by Watson Crick base pairing or Hoogsteen G-G- base pairing. On using 7-deaza-2'-deoxyguanosine-5'-triphosphate (C^7dGTP), there is preclusion of Hoogsteen base pairing and eliminates the formation of guanine mediated aggregations, thus relieves band compression. Similarly, use of C^7dATP precludes the band compression due to secondary structures formed due to dATP.

7.7. Automated Sequencing

In manual sequencing, the DNA fragments are radio-labeled in four chain termination reactions, separated on the sequencing gel in four lanes, and detected by autoradiography. This approach is not well suited to automation. To automate the process, it is desirable to acquire sequence data in real-time by detecting the DNA bands, within the gel, during the electrophoretic separation. Detecting the bands within the gel is not trivial, as there are only about 10^{-15} to 10^{-16} moles of DNA in each band. The solution to the detection problem is to use fluorescence methods.

In the first generation of fluorescence techniques, fluorescent tags were attached to the sequencing primer. By attachment of the fluorescent dyes to the primer, each of which can be discriminated by its fluorescent spectrum, it is possible to electrophorese the products of all four chain terminator reactions together in one lane of a sequencing gel. The DNA bands are detected by their fluorescence as they pass the detector after electrophoresis. If the detector is made to scan horizontally across the base of a slab gel, many separated sequences can be scanned as one sequence per lane.

7.8. Analysis of DNA Sequence Data by Computer (in Silico)

Any substantial DNA sequence undertaking will necessitate computer analysis of the data. The analysis will often fall in two stages. In the first stage, compiling the data; computer files will be more or less directly fed with sequence information read from the autoradiograph of the gel or other sequencing system. Automatic system has the obvious advantage of speed and, just as importantly, they help to minimize the ever present difficulty of clerical errors accumulating in sequence data. The compiler will then be used to compile readings from several sequencing runs, to search for and identify overlaps and hence *contiguities* between runs, and to compile data from sequencing complementary strands of the DNA. The software will also identify inconsistencies indicating sequencing or complementary strands.

A sequencing gel is a high resolution gel designed to fractionate single stranded (denatured) DNA fragments on the basis of their length. Such gels routinely contain 6 to 20% polyacrylamide and 7.8 M urea. The urea is a denaturant whose function is to minimize DNA secondary structure. The gel's temperature is also maintained nearly 55°C (minimum 50°C and maximum 65°C) in order to keep DNA in denatured form and to avoid secondary structures.

The Procedure

The DNA to be prepared as a single strand. This template DNA is supplied with:

- A mixture of all four normal (deoxy) nucleoside triphosphates in ample quantities.

  *dATP
  *dGTP
  *dCTP
  *dTTP

- A mixture of all four dideoxynucleoside triphosphates, each present in limiting quantity and each labeled with a 'tag' that fluoresces a different color:

*ddATP
*ddGTP
*ddCTP
*ddTTP

- DNA polymerase I

In the presence of all the four DNA normal nucleotides, chain elongation proceeds normally until, by chance, DNA polymerase inserts a dideoxy nucleotide instead of normal deoxynucleotide. If the ratio of normal nucleotide to the dideoxy version is high enough, some DNA strands will succeed in adding several hundred nucleotides before insertion of the dideoxy versions halts the process.

At the end of the incubation period, the fragments are separated by length from longest to shortest. The resolution is so good that a difference of one nucleotide is enough to separate that strand from the next shorter and the next longer strand. Each of the four dideoxynucleotides fluoresces a different color when illuminated by a laser beam and an automatic scanner provides a print out of the sequence.

7.9. RNA Sequencing

Like DNA, RNA can be sequenced by enzymatic synthesis. A suitable primer (which therefore requires some preliminary sequence information to be available) is annealed at the 5' end of the RNA. The sequence reactions are carried out as with DNA sequencing except that reverse transcriptase is used instead of DNA polymerase I. The product of the four reactions are analyzed electrophoretically exactly as before.

References

[1] Maxam, A.M. and Gilbert, W. (1977) A new method for sequencing DNA. *Proc. Natl. Acad. Sci. USA* **742:** 560-564.
[2] Maxam, A.M. and Gilbert, W. (1980) Sequencing end labeled DNA with base specific chemical cleavage. *Methods Enzymol.* **65:** 499-560.
[3] Sanger, F., Nicklen, S. and Coulson, A.R. (1977) DNA sequencing with chain terminating inhibitors. *Proc. Natl. Acad. Sci. USA.* **74:** 5463-5467.
[4] Holley, R.W., Marquisse, M., Merrill, S.H., Penswick, J.R. and Zamir, A. (1965) Structure of a ribonucleic acid. *Science* **147:** 1462-1465.
[5] Sanger, F. and Coulson, A.R. (1975) A rapid method for determining sequences in DNA by primed synthesis with DNA polymerase. *J. Mol. Biol.* **94:** 444-448.
[6] *Analects* (1991) Vol. 19(4) (A Pharmacia PL Biochemicals News Letter) New ! 7-Deaza-2'-deoxyadenosine 5'-triphosphate (c^7dATP).

Chapter VIII

Oligonucleotide Site Directed Mutagenesis Techniques

8.1. Introduction

Oligonucleotide site directed mutagenesis technique is the real genetic engineering tool with which one can make a desired point mutation in a gene. However, the prior knowledge of the complete nucleotide sequence of the gene is essential. The technique is mainly used for determining the importance of a particular point mutation which is done to change a particular amino acid in a protein chain. Therefore, the technique became popular for determining the amino acid in an enzyme protein to be involved in enzyme catalysis or essential for proper activity of the enzyme. It was tried to use the technique in manipulating the enzyme proteins for getting more active enzyme. However, the main problem is that we do not know which amino acid should be changed for getting more active enzyme protein. In creating a desired mutation in a gene, help of an oligonucleotide is taken which is designed in such a way that it should have mismatch in the middle of the oligonucleotide and it is annealed at the complementary site. It is ensured that it should get annealed at the desired site only and not on any other site. Afterwards, it acts as a primer and synthesis of the complementary strand is carried out with the help of the Klenow fragment. The original method for site directed mutagenesis was given by Zoller and Smith. Afterwards, certain improvements were carried out for making the technique more effective.

8.2 Zoller & Smith Procedure for Site Directed Mutagenesis

The basic principle of the technique is that the gene in which mutation has to be carried out is cloned in a M13mp vector using a restriction endonuclease recognition site present within the multiple cloning site region. The recombinant double stranded (replicative) form of M13mp vector is transformed in a suitable strain of *E.coli* like JM 89 or JM 101 or 109 using routine procedure of transformation. After transformation, replication of the M13mp vector DNA is carried out and as a result, there is synthesis of the positive strand of M13mp DNA which after synthesizing the viral coat, comes out of the bacterial cell without rupturing the cell. These viral particles are isolated and recombinant M13mp DNA (positive strand) is isolated (as described in Chapter 3 under vectors). On the other hand, a suitable primer is designed at the site of creating the mismatch. Care is taken that mismatch should be in the middle of the primer so that it should not get chewed by the exonuclease activity of the DNA polymerase. On an average, the size of the primer is 16 to 20 mers. It should get annealed at the desired site only. If in a trial experiment, it is observed that it is also annealing at other site(s) too, then the size of the primer may be increased. Sometimes for rapid replication, extra (one or two) primers are also designed having completely complementary sequences which bind at other site(s) away from the site of interest (where a mismatch containing primer is annealed). Mostly, these extra primers are annealed in the M13mp vector sequence. Under the conditions, replication occurs from more than one end. After annealing, the template (positive recombinant strand) having primer(s) is added in the system containing all the components required for DNA replication including all the four deoxynucleoside triphosphates, Klenow fragment and DNA ligase (for nick closure). Following synthesis of the complementary strand, either the newly synthesized strand is separated using suitable technique or the mixture as such is transformed in a suitable bacterial cell (*E.coli*). Screening is done for selecting the phage DNA having desired mismatch in the sequence by using primer having mismatch as a probe. For the purpose, the primer is phosphorylated using radioactive phosphorus with the help of polynucleotide kinase. After screening, the entire sequence is determined to confirm the point mutation at the desired site. Once, desired mutation is there, the mutant gene may be cloned in a suitable expression vector and mutant protein may be isolated and characterized.

Primer design: The oligonucleotide is designed so that mismatch(s) is located near the middle of the oligonucleotide. It is important especially when the oligonucleotide is used as a probe to screen for the mutant. If the mismatch is in the middle, it yields the more binding differential between a perfectly matched strand (wild type) and a mismatch containing strand (mutant). Besides, presence of the mismatch in the middle protects it from exonuclease activities of the DNA polymerase. *Escherichia coli* DNA polymerase I has both 5'-3' and 3'-5' exonuclease activities. The Klenow fragment which is a preferred enzyme in site directed mutagenesis has 3'-5' exonuclease activity. If the mismatch is placed near the 3' end, it may result in the repair of the mismatch by the 3'-5' exonuclease activity of the Klenow fragment. Although, it has been shown that three nucleotides following the mismatch are enough to protect against the exonuclease activity. The other important point is to carry a computer analysis on the DNA sequence of the vector and the cloned fragment for regions of partial complementarity. If complementary sequence is present at other site(s), size of the oligonucleotide should be extended.

Synthesis of the oligonucleotide: Although both chemical and enzymatic procedures are available for synthesis of the oligonucleotide, chemical procedures are preferred since enzymatic methods give relatively lower yield of the final product. The chemical procedures are the methods of choice. Mostly, methods based on phosphoramidite, phosphotriester or phosphite chemistry are used. Nowadays, automatic computerized DNA synthesizers are available which use solid phase procedures for synthesis. After synthesis of the oligonucleotide which is obtained in nanomolar quantities are further purified using ion exchange and reverse phase chromatographies. Once the oligonucleotide is synthesized, it should be sequenced to confirm the desired sequence. Mostly chemical method of Maxam and Gilbert with modifications is used for sequencing the small oligonucleotides. For the purpose, the oligonucleotide is first 5' end phosphorylated and end labeled using γ-^{32}P ATP and polynucleotide kinase reaction.

Protocol for End Labeling: To an eppendorf tube add 20 microcurie of γ-^{32}P ATP and dry in a speed vac and add 20-50 picomoles of the oligonucleotide, add 3 microlitre of a buffer containing 1M tris-HCl, pH 8.0, 1.5 μl of 0.2M magnesium chloride, 1.5 μl of 0.1 M dithiothreitol (DTT) and water to make total volume 30 μl. The contents are mixed properly taking all the possible care for radioactivity. Add about 5 units of T4 polynucleotide kinase, mix and incubate at 37 °C for about an hour. Stop the reaction by heating at 65°C for 10 minutes. After that, the contents are chromatographed onto a Sephadex G-25 column to remove unincorporated radio active ATP whose peak is eluted after the peak of the oligonucleotide labeled at 5' end. The fractions are analyzed on the basis of radioactivity. For Sephadex G-25, column should be made of either plastic or siliconized glass. 50 mM ammonium bicarbonate, pH 7.8 may be used as elution medium. One may collect the fractions of 100 μl each. The major fractions containing radioactive oligonucleotide are pooled. The radioactive oligonucleotide is stored at –20°C since, besides sequencing the oligonucleotide, it is also used as probe for screening the mutant. With the protocol, generally 2 to 6 x 10^6 cpm per 20 pmole are incorporated. In addition to *in vitro* mutagenesis experiment, the oligonucleotide is also phosphorylated at 5' end using cold ATP (nonradioactive) and T4 polynucleotide kinase. The basic protocol is same as described above for getting radioactive labeled end.

Protocol for Sequence Determination of the Oligonucleotide

1. Take about 500,000 cpm of the radioactive labeled oligonucleotide in a siliconized eppendorf tube and dry in a speed vac. Resuspend it in 33 μl of water. It is not always necessary to dry the sample. If 500,000 cpm are present in 1 microlitre, 32 microlitre water may be added in it to make 33 microlitre.
2. Add 5 microlitre in G tube, 6 microlitre in C tube, 10 microlitre in G + A tube and 11 microlitre in C + T tube. Afterwards, add 2 microlitre of 1 mg/ml denatured calf thymus DNA in each tube.
3. Add 20 μl of 5M NaCl to the C tube and 100 μl of cacodylate buffer to the G tube.
4. Take 300 μl of cacodylate buffer in a separate tube and add 1 μl of dimethyl sulfate in it.
5. To start the reaction, add 100 μl of the diluted dimethyl sulfate in G tube (really we are adding 1/3rd microlitre of dimethyl sulfate in it and incubate at 20°C for 3 minutes.
6. Add 3 μl of 50% formic acid in the G + A tube and incubate at 37°C for 10 minutes.

7. Add 30 μl of hydrazine in each C and T + C tubes. Mix and incubate at 20°C for 3 minutes.
8. Stop reaction in G tube by adding 50 μl of DMS stop. In other tubes, the reaction is stopped by adding HZ stop (220 μl in G + A tube and 200 μl each in C and T + C tubes). Immediately add 750 μl of chilled 95% alcohol in all the tubes and chill all the tubes in dry ice ethanol bath for 10 minutes. Centrifuge for 10 minutes in a microfuge and remove all the supernatant carefully without disturbing the pellet.
9. Follow all the other steps further as described in Chapter VII.
10. Electrophorese the modified samples on a 20% polyacrylamide urea gel until the bromophenol blue is approximately 1/3rd down the gel. Develop the bands using autoradiography.

- **DMS buffer:** 50 mM sodium cacodylate buffer, pH 8.0 containing 1 mM EDTA
- **DMS stop:** 1.5 M sodium acetate, pH 7.0 containing 1 M 2-mercaptoethanol and 100 μg/ml tRNA.
- **Hydrazine (HZ) stop:** 0.3 M sodium acetate containing 0.1 mM EDTA and 25 μg/ml tRNA
- **Loading buffer:** 0.1% bromophenol blue + 0.1% xylene cyanole in 100% deionized formamide.
- **For Siliconization** of glass plates or tubes etc,, 1% (V/V) dimethyl dichloro silane in chloroform or carbon tetra chloride.

Preliminary *in Vitro* Tests for Specific Priming

a. Primer extension: Add the following in a siliconized eppendorf tube:

1. M13 recombinant DNA (approx. 0.5 pmole)
2. 5' end labeled oligonucleotide (nearly 10-20x molar excess over the template
3. 1 μl of solution A (0.2M tris-HCl, pH 7.5, 0.1 M $MgCl_2$, 0.5 M NaCl, 0.01 M DTT) and water to adjust 10 μl.
4. Heat at 55°C for 5 minutes using a water bath.
5. Cool to room temperature, and add 1 μl of 2.5 mM dCTP, 1 μl of 2.5 mM dTTP, 1 μl of 2.5 mM dGTP, 1 μl of 2.5 mM dATP, 0.5 μl of solution A and mix.
6. Add 1-2 units of Klenow fragment.
7. Mix and incubate at 25°C for 5 minutes.
8. Stop the reaction by incubating at 65°C for 10 minutes.
9. Cool to room temperature and adjust buffer for appropriate restriction endonuclease
10. Add 2-5 units of the restriction endonuclease and incubate at the suitable temperature for one hour
11. Add equal volume of formamide dye solution
12. Electrophorese the sample on a 6% polyacrylamide gel under denaturing conditions.
13. Autoradiograph to detect the expected restriction fragments

b. Chain terminator sequencing: Use the procedure of Sanger et al (dideoxy method) for chain terminator sequencing. Substitute the mismatched oligonucleotide for a universal M13 sequencing primer.

1. Add to a siliconized tube: M13 recombinant DNA (0.1- 0.5 pmole), oligonucleotide (10-30x over template DNA), 1 μl of solution A and water to make 10 μl
2. Mix and close the lid, heat at 55°C for 5 minutes, cool to room temperature and follow Sanger's procedure of sequencing.

Using any of the above methods, one may detect whether the primer has annealed at the desired place only. Once it is confirmed, the actual mutagenesis work is carried out.

Mutagenesis

a. Annealing

1. Add M13 recombinant DNA (nearly 1 pmole), 5'-phosphorylated oligonucleotide (10-30x molar excess over the template DNA), 1 μl of solution A and water to make the volume 10 μl in an eppendorf tube.
2. Mix the contents and incubate in a water bath at 55°C for 5 minutes for complete denaturation. Afterwards, incubate the tube at the room temperature (25°C) for 5 minutes.

The above steps will anneal the oligonucleotide on the complementary site (site of interest where in the middle is the mismatch with the oligonucleotide).

b. Extension and ligation

1. Prepare solution C (1 μl of solution B {0.2M tris-HCl, pH 7.5 containing 0.1 M $MgCl_2$ and 0.1 M DTT}, 1 μl each of 10 mM dCTP, dTTP, dGTP, 0.5 μl of dATP, 1 μl of 10 mM ATP, 1.5 μl of α-^{32}P-dATP {10 millicurie/ml}, 2 units of T4 DNA ligase, 2 μl of water). Prepare fresh before adding to the annealed DNA. Keep on ice before use.
2. Add 10 μl of Solution C to the annealed DNA.
3. Mix thoroughly and add 2-3 units of the Klenow fragment, mix and incubate at the room temperature for 10 minutes.
4. Add 1 μl of 10 mM dATP
5. Mix and incubate at 15°C overnight.

The above protocol is for 15-20 mers oligonucleotide. For shorter oligonucleotides, initial extension reaction is carried out at 10°C and afterwards continued at 15°C overnight. As a result, there is extension of the complementary strand and nick(s) are closed with the help of DNA ligase. This procedure synthesizes a double stranded circular DNA from single stranded DNA.

c. PEG-NaCl precipitation

This step is carried out to remove unincorporated radioactive dATP and also for precipitation of DNA.

1. To the reaction vessel, add 30 μl of water and 50 μl of PEG-NaCl (1.6 M NaCl and 13% PEG of the molecular weight 6000).
2. Mix and incubate in an ice bath for 15-20 minutes. The step brings out precipitation of DNA.
3. Centrifuge in a microfuge at the room temperature for 5-10 minutes, and remove all the supernatant.
4. Add 100 μl of chilled 0.8 M NaCl-6.5% PEG-6000 to rinse the walls of the tube. Suspension of the pellet in the PEG-NaCl solution is not done but it is done just to remove any adhering impurity. Centrifuge again in the microfuge at the room temperature for 10 minutes and remove the supernatant completely.
5. Suspend the pellet in 100 μl of TE buffer.

One may check the amount of radioactivity (cpm) in the solution after step 5 to evaluate the extension process. At this stage, it has double stranded closed circular DNA, closed circular single stranded DNA, open circle DNA. Some people separate double stranded closed circular DNA from other DNAs using alkaline sucrose gradient centrifugation. However, others do not prefer to purify double stranded closed circular DNA because the amount is so small that there are chances of loss during density gradient centrifugation.

d. Alkaline sucrose gradient centrifugation

Prepare a sucrose density gradient in a centrifuge tube using 5 to 20% sucrose. For the purpose, starting from the 20 % sucrose one ml of each of the sucrose solution is layered carefully to avoid mixing. Store the tube in the cold overnight (8-16 hours) to linearize the gradient. For shortening, the tube may also be left at the room temperature for 2 hours followed by leaving it in the cold condition (0-4°C) for 2 hours. After the gradient formation, sample preparation is done as follows:

1. Add 20 μl of 2N NaOH to the sample after PEG-NaCl precipitation.
2. Mix, incubate at the room temperarure for 5 minutes followed by incubation on ice for one minute.
3. Put the gradient tube into the SW 50.1 rotor and afterwards, layer the sample on the top of the gradient carefully
4. Centrifuge at 40,000 rpm for 2 hours in the cold condition. After centrifugation, collect fractions (0.5 –1 ml) in eppendorf tubes after puncturing the tube at the bottom without disturbing the contents. DNA containing fractions may be screened taking the advantage of the radioactivity in the DNA. The double stranded closed circular DNA remains in the bottom half of the gradient, whereas, single stranded circular and open circle DNA molecules remain in the upper half of the gradient.

Pool the closed circular double stranded DNA fractions and neutralize using 1 M tris-citrate, pH 5.0. On an average, 50 μl of tris-citrate is added per 300 μl of the solution. Amount of the radioactivity in the sample may be measured in a scintillation counter to calculate the percentage of closed circular double stranded DNA formation.

Now as written above, either the sample without purification through density gradient or the purified closed circular double stranded DNA is used for transformation in a suitable *E.coli* strain like JM101.

Transformation

a. Competent cells preparation

1. Inoculate 2 ml of 2xYT medium (1xYT is 8g tryptone, 5g yeast extract and 8 g NaCl in one litre, adjusted to pH 7.6) with a single colony of JM101. Incubate in an environmental shaker at 37°C, 200 rpm overnight.
2. Add 0.3 ml of overnight culture to 30 ml of 2xYT medium, incubate at 37°C with shaking at 200 rpm till OD_{600} is nearly 0.4.
3. Harvest the cells by centrifugation in cold condition at 10,000 rpm for 5 minutes.
4. Wash the cells with ½ volume (15 ml) of cold 10 mM NaCl.
5. Suspend the pellet in ½ volume (15 ml) of 30 mM $CaCl_2$ and let stand on ice for 20 to 30 minutes.
6. Centrifuge at 5000 rpm for 10 minutes in the cold condition and gently resuspend cells in $1/10^{th}$ volume of the cold 30 mM $CaCl_2$ containing 15% glycerol. After calcium chloride treatment which makes the cells competent, the cell wall of the cells becomes fragile, therefore, care must be taken while handling.
7. Store in 200 μl aliquots at –70 to –90°C. These cells may be used upto two weeks. Although the efficiency of the competence decreases on storage. According to some people, upto 24 hours efficiency increases and thereafter decreases.

It is considered that glycerol decreases the competence of the cells. It is added only for storing the cell without rupturing at –70°C. If anyone has to use immediately after the preparation, one may suspend only in calcium chloride solution.

b. Transformation of competent cells with closed circular double stranded DNA

1. Keep the 200 μl aliquot of the competent cells on ice. If it has been taken out from the deep freezer, it should be allowed to thaw on ice.
2. Add a suitable aliquot of the closed circular double stranded DNA in it. As written above, DNA sample after PEG-NaCl precipitation without purification may also be used for transformation. Amount of the DNA should be nearly 100 nanogram. It should not be more than a microgram. It is preferred to prepare controls too. In one tube, no DNA should be added and to another double stranded M13 DNA (with no recombinant) should be added.
3. Incubate the tubes on ice for 50-60 minutes.

4. Heat shock the tubes by keeping in a water bath adjusted at 42°C for 60 to 120 seconds and immediately transfer back on ice. It is preferred to incubate at 42°C for 90 seconds.
5. Add 3 ml of soft agar in YT (0.75% agar in YT medium) and 200 µl of the lawn culture (JM101 exponential phase culture) to each tube, vortex gently to mix and pour onto a YT plate.
6. Allow to harden the soft agar at the room temperature for about 30 minutes, thereafter incubate at 37°C for 8 to 10 hours (overnight). Plaques appear on the plate.

The number of plaques varies from plate to plate ranging 10 to 250 plaques per nanogram of the closed circular double stranded DNA (CC-DNA) added depending on the efficiency of the competent cells. In the control, the number of plaques may be even 1000 or more. If no plaque appears from the CC-DNA, dialysis of the fraction is carried out against one litre of 2 mM tris-HCl, pH 8.0 containing 0.01 mM EDTA and transform again. If still no plaque is there, the sample of DNA may be lyophilized for concentration.

Isolation of the Phage

Prepare phage cultures from individual plaques using the following protocol. Depending on the person's efficiency, 32 to 48 individual plaques may be handled at one time.

1. Dilute the overnight JM101 culture 50 times with 2x YT medium.
2. Incubate with shaking for 2 hours at 37°C.
3. Dilute 10 x to this culture with 2x YT medium and transfer 2 ml to each of the culture tubes number equal to the number of the plaques decided to be handled at a time.
4. Suck each phage plaque from the plate with the help of a glass disposable Pasteur pipette and blow alongwith the agar plug into the culture tube.
5. Incubate the tubes in an environmental shaker at 37°C, 200 rpm for 6 hours. During the incubation, the phage infects the bacterial cells, multiply and after multiplication comes out in the medium without rupturing the cells. Longer incubation is not recommended since on longer incubation, there are chances of rupturing the bacterial cells.
6. Transfer the contents (1.5 ml) of each tube into 1.5 ml eppendorf tubes. Spin the eppendorf tubes in a microfuge for 5 minutes at the room temperature.
7. Pour the phage particles containing supernatant into fresh sterilized eppendorf tube.
8. Add 300 µl (20 % of the volume) of 20% PEG-2.5 M NaCl to each tube.
9. Mix thoroughly by vortexing and incubate at the room temperature for 15 minutes. This allows the precipitation of the phage particles.
10. Centrifuge in a microfuge for 10 minutes at the room temperature and afterwards, remove the supernatant carefully. The traces of the supernatant may also be removed with the help of the micropipette. The pellet contains the phage particles.
11. Suspend the phage pellet in 100 µl of TE buffer.

12. For doing hybridization screening, the protocol may be stopped at this stage and phage particles may be used directly for hybridization studies. For sequencing or restriction endonuclease site screening, the protocol is continued further for isolation of the phage DNA.
13. Add equal volume of the phenol chloroform isoamylalcohol mixture to denature and extract the proteins (viral coat proteins).
14. Vortex for few seconds, centrifuge for 2 minutes at the room temperature. The centrifugation separates the layers well.
15. Remove the upper aqueous phase carefully and transfer into an another eppendorf tube.
16. Repeat steps 13-15 at least once. It is preferred to repeat twice.
17. Extract the aqueous phase with 0.5 ml of water saturated diethyl ether.
18. Repeat step 17 at least once.
19. Incubate the aqueous phase at 37°C in a water bath for few minutes to evaporate the traces of diethyl ether, if any.
20. Add 3 M sodium acetate, pH 4.8-5.0 to get final concentration of 0.25 M. Afterwards, add three volumes of the chilled 95% ethanol.
21. Precipitate DNA by chilling the tubes in dry ice ethanol bath for 15-20 minutes.
22. Centrifuge the tubes in a microfuge in the cold condition for 10 minutes to settle the precipitated phage DNA.
23. Remove the supernatant carefully without disturbing the phage pellet. The traces of the supernatant may be removed with the help of a micropipette.
24. Wash the pellet with chilled ethanol and remove the supernatant after centrifugation.
25. Dry under vacuum (Speed vac or equivalent may be used).
26. Suspend the phage DNA in 20 μl of TE buffer.

Screening Procedures

One may decide which screening procedure is suitable for screening the mutant. The mutation may be detected by sequencing with a M13 sequencing primer, another synthetic oligonucleotide or a restriction fragment. Restriction site screening may be used if the mutation creates or destroys a restriction endonuclease recognition site. The hybridization method is simple and commonly used.

a. Chain termination sequencing: One may use 2-3 μl phage DNA as template with only one dideoxyribonucleotide in a chain terminator reaction. The primer may be an universal M13 primer, another synthetic primer (oligonucleotide) or a restriction fragment. Suspected mutants are sequenced using all the four dideoxyribonucleotides mixture.

b. Restriction site screening

1. Add 2-3 μl of the phage DNA in a siliconized eppendorf tube and afterwards add 1 μl of solution A, 1.5 pmole of M13 universal sequencing primer and water to adjust the volume 10 μl.

2. Mix thoroughly and incubate at 55°C for 5 minutes. Thereafter, cool to the room temperature slowly. If a restriction fragment is used as a primer, heat the mixture for 5 minutes at 100°C, hybridize at 67°C for 30 minutes and thereafter cool to the room temperature.
3. Add 1 μl of each of 0.5 mM solutions of dTTP, dCTP, and dGTP; 0.5 μl of 0.1 mM dATP, 5 μCi of (α-^{32}P) dATP, 0.5 μl of solution A and mix properly.
4. Add 1 unit of Klenow fragment, mix and incubate for 10 minutes at the room temperature.
5. Add 1 μl of 2.5 mM dNTPs, mix and incubate for 5 minutes at room temperature.
6. Heat at 65°C for 10 minutes, cool to room temperature.
7. Adjust the buffer for desired restriction endonuclease.
8. Add 2-5 units of the restriction endonuclease, mix and incubate for one hour at the appropriate temperature.
9. Stop the reaction by adding 2 μl of the dye (0.1% bromophenol blue, 0.1% xylene cyanol, 15% glycerol) solution.
10. Load the sample onto a 5% non-denaturating polyacrylamide gel or 1% agarose gel depending upon the size of the fragments and electrophorese.
11. Autoradiograph

c. Dot blot hybridization

1. Spot 2 μl of phage particle suspension onto a dry sheet of nitrocellulose.
2. Air dry and bake at 80°C for 2 hours in a vacuum oven.
3. Prehybridize with 6x SSC (stock solution 20x SSC is 3 M NaCl, 0.3 M sodium citrate, 0.01 M EDTA, adjusted to pH 7.2) + 10x Denhardt's solution (stock solution 100x Denhardt's solution is 2% bovine serum albumin, 2% PVP, 2% ficoll) + 0.2% SDS at 67°C.
4. Remove the prehybridization solution and wash the nitrocellulose membrane gently with 50 ml of 6x SSC for 1 minute at 23°C.
5. Place the nitrocellulose membrane into a sealable bag, add DNA probe and seal. The probe is 5' end radioactive labeled oligonucleotide in 6x SSC + 10x Denhardt's solution (without SDS). On an average, 10^6 cpm are added. The membrane should remain dipped in the solution.
6. Incubate at the room temperature (25°C) for one hour.
7. Remove the membrane from the bag, wash in 6x SSC at the room temperature. One may use 3X 50 ml 6x SSC for wash for a total time of 10 minutes.
8. Dry by keeping between the paper towels and autoradiograph for one hour at the room temperature. One may also do at –90°C to shorten the exposure time.
9. Wash the membrane at a higher temperature with 50 ml of 6x SSC for 5 minutes. The wash solution should be preheated to the desired temperature.
10. Autoradiograph in the same way as mentioned in step 8.
11. The steps 9 and 10 are repeated using several wash temperatures, everytime increasing by 5°C.

The dot blot hybridization may also be carried out with phage DNA. A specific apparatus of dot blot hybridization available commercially may be used but there is no absolute need for that apparatus and it can be done manually.

The choice of the temperature is dependent on the length and GC % of the oligonucleotide. The specific mutation constructed contributes to the ability to distinguish between wild type and mutant molecules since mutant will be 100% complementary whereas wild type will have a mismatch. Although mostly the protocol has been used for constructing single point mutants, it may also be used for constructing double point mutants.

Although theoretically, there should be 50% mutants at the time of screening the phage particles. Practically, the number of mutants has been found to be very low perhaps due to the repairing mechanism of the enzyme/cell. Later on the protocols have been modified to get more number of mutant molecules.

8.3. Kunkel Procedure for Site Directed Mutagenesis

As mentioned above, due to the practical reason of getting much lesser number of mutants, modifications in the protocol have been suggested. One such modification was carried out by Kunkel. There are number of features which make this method of choice. This method requires lesser number of experimental steps and enzymes. The basic principle of the method is as follows:

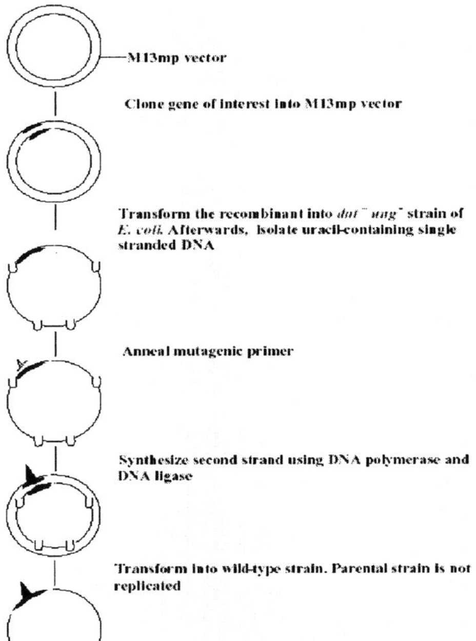

Figure 8.1. Steps involved in *in vitro* mutagenesis using Kunkel method of site directed mutagenesis

The gene of interest is cloned in M13 vector (or in any other suitable vector capable of producing single stranded DNA). The recombinant M13 vector is transformed in *E.coli* strain having *dut*⁻ (*dut* is the gene coding for dUTPase enzyme which hydrolyzes dUTP) and *ung*⁻ (*ung* is the gene coding for the enzyme uracil N-glycosylase which removes uracil from DNA). After transformation, recombinant vector is allowed to multiply within the bacterial cell. Since the *E.coli* cell is deficient in dUTPase, there will be increased level of dUTP which will compete for dTTP during DNA replication resulting in incorporation of dUTP in DNA and the cell is also deficient in uracil N-glycosylase which removes uracil from DNA, the incorporated dUTP will remain in DNA and is called uridylated DNA. Single stranded uridylated recombinant phage DNA is isolated and used as a template. The designed oligonucleotide having mismatch nearly in the middle of the oligonucleotide is annealed with the template and *in vitro* mutagenesis experiment (replication of the complementary strand) is carried out. Afterwards, closed circular double stranded DNA which has one strand uridylated and the other as normal strand (having thymine nucleotide) is transformed into wild type strain of *E.coli*. Here, due to uracil N-glycosylase enzymes, uridylated DNA gets chewed. Only the thymidylated DNA is stable (Figure 8.1). Thus, in this method, there are more chances of getting mutant molecules. Nowadays, many commercial companies sell kits having the required reagents, enzymes and bacterial strains. The practical details of the method are similar to as described above for Zoller and Smith procedure.

8.4. Gapped Duplex Procedure for Site Directed Mutagenesis

The gapped duplex procedure for site directed mutagenesis was described by Kramer *et al.* It involves the binding of the oligonucleotide to gapped duplex M13 vector DNA. That's why, it is called as gapped duplex method. The principle of the method is as follows:

Gapped duplex DNA formed by annealing double stranded DNA to circular single stranded DNA having an insert in it. The double stranded M13 phage vector is linearized using the restriction endonuclease used for cloning the insert. After annealing of the double stranded DNA to circular single stranded DNA having insert results in the formation of a gapped duplex having double strand at the portion derived from the M13 vector and single stranded at the insert. The desired mutation is constructed with an oligonucleotide which has a single mismatch nearly in the middle and is hybridized at the single strand portion of the negative strand. The rest of the portion of the negative strand is synthesized using Klenow fragment and DNA ligase. This double stranded DNA is transformed into an *E. coli* strain which lacks a mismatch repair system. Therefore, the mutation formed by the oligonucleotide remains coupled to the negative strand. The mutated phage can be specifically selected. It is possible to distinguish between phage derived from positive and negative strands because the phage having the original insert also contains amber mutations in essential genes. Amber mutations are lacking in the negative strand. If the phage is transformed in an amber suppressor host, amber mutations will be rescued. If the E.coli is a suppressor negative strain, only the phage without amber mutations will grow.Therefore, on transformation in a suppressor negative strain, only negative strand which also has mismatch (mutation) will be multiplied (Figure 8.2). This method may give even upto 90% amplification of the desired mutation. Nowadays, commercial companies sell kits having reagents, enzymes and required bacterial strains for carrying gapped duplex method for site directed mutagenesis. The practical details are mainly same as written above for Zoller and Smith procedure.

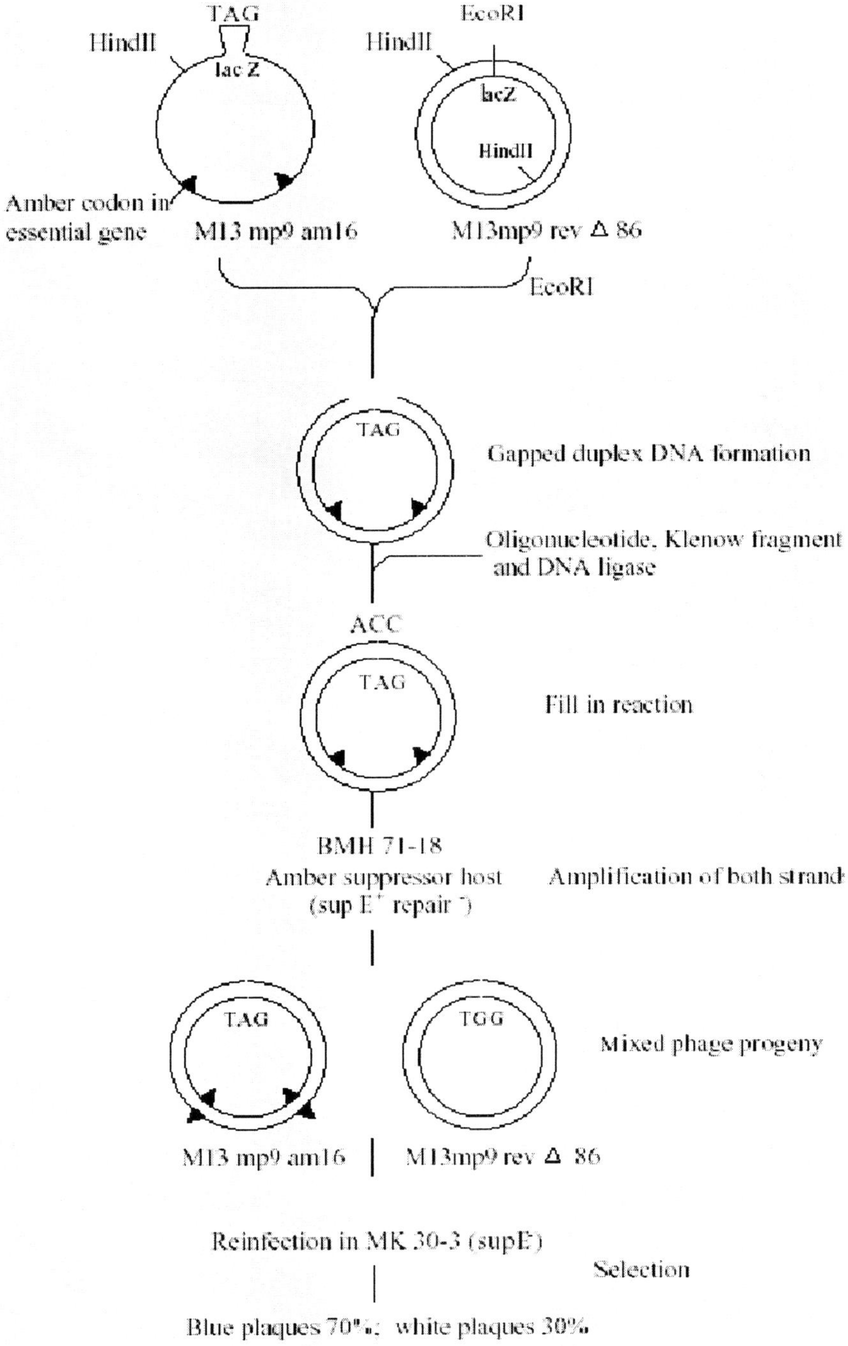

Figure 8.2. Steps involved in *in vitro* mutagenesis using gapped duplex method

Other procedures of site directed mutagenesis are also in the literature, all are mainly based on the principle that an asymmetry is constructed between the two strands of the heteroduplex, allowing the selection between the mutant and the wild type. Some of the other methods are use of hemimethylated DNA to take advantage of the preferential repair of the methylated strand; use of phosphorothioate modified DNA; double primer method using the conversion between the EcoK and EcoB restriction modification systems. But these methods are not in common use.

References

[1] Zoller, M. and Smith, M. (1983) Oligonucleotide directed mutagenesis of DNA fragments cloned into M13 vectors. *Methods Enzymol.* **100:** 468-500.

[2] Kunkel, T.A., Roberts, J.D. and Zakour, R.A. (1987) Rapid and efficient site specific mutagenesis without phenotypic selection. *Methods Enzymol.* **154:** 367-382.

[3] Geisselsoder, J., Witney, F. and Yuckenberg, P. (1987) Efficient Site-Directed In Vitro Mutagenesis. *Biotechniques* **5:** 786-791.

[4] Kramer, W., Drutsa, V., Jansen, H.W., Kramer, B., Pflugfelder, M. and Fritz, H.J. (1984) The gapped duplex DNA approach to oligonucleotide-directed mutation construction. *Nucleic Acids Res.* **12:** 9441-9458.

[5] Marmenout, A., Remaut, E., VanBoom, J. and Fiers, W. (1984) Oligonucleotide directed mutagenesis: selection of mutants by hemimethylation of GATC sequences. *Mol. Gen. Genet.* **195:** 126-133.

[6] Taylor, J.W., Ott, J. and Eckstein, F. (1985) The rapid generation of oligonucleotide directed mutations at high frequency using phosphorothioate modified DNA. *Nucleic Acids Res.* **13:** 8765-8785.

[7] Carter, P., Bedouelle, H. and Winter, G. (1985) Improved oilgonucleotide site directed mutagenesis using M13 vectors. *Nucleic Acids Res.* **13:** 4431-4443.

Chapter IX

Polymerase Chain Reaction

9.1. Introduction

Polymerase chain reaction (PCR) is a technique sensitive enough to amplify small DNA fragments to a billion folds. The generation of amplified fragments (called as amplicons) using conventional PCR or by the modified methods of PCR are extensively used to study a large number of queries related to biology, medicine and forensic sciences. With the help of PCR, many genetic diseases may be diagnosed that are caused by protozoa, parasites, bacteria, viruses and monitored with ease and accuracy.

PCR utilizes the principle of DNA replication. DNA polymerase uses the single stranded DNA as a template for the synthesis of a complementary new strand. It requires a small oligo-as primer to initiate the synthesis of the complementary strand. Therefore, the starting point of DNA synthesis depends upon an oligonucleotide primer that anneals to the template at the complementary site. In PCR, both the strands of a double stranded DNA may act as templates if the primer is supplied for both the strands. For PCR, the primers are selected to flank the regions of DNA that is to be amplified so that newly synthesized strands of DNA, starting at each primer extend beyond the position of the primer on the opposite strand. Thus new primer sites are generated on each newly synthesized DNA strand. Generation of amplicons with a pair of primers is usually known as PCR or symmetrical PCR.

9.2. Constituents of PCR

Laboratory conditions: The most important aspect for successful PCR run is the PCR laboratory. Conditions of the isolated PCR laboratory are important to avoid contamination of the new PCR assays with the old PCR products. Due to much higher sensitivity of PCR, a clean DNA free environment is needed and care is taken that only DNA of interest should enter in the PCR reaction. The laboratory should be comparable to sterile microbiological environment as biohazard infects the PCR and not the researcher. Two different types of PCR laboratories are established depending upon the requirement: (1). Contamination sensitive-used for quantitative assessment or when qualitative detection of large number of samples is carried out. (2). Contamination insensitive- if it is used for mutagenesis of a DNA clone.

The following approaches are applied to prevent PCR contamination:

1. Physical separation is done between Pre PCR (sample preparation), PCR and post PCR (analysis) areas.
2. Use of ultraviolet (UV) lights is done. UV lights reduces contamination levels by several logs. However, UV light is less effective if the DNA is of less than 300 base pairs.
3. Sometimes, use of uracil DNA-glycosylase and dUTP is done in place of dTTP.

The ideal PCR laboratory should have the following distinct areas well separated from each other:

(a) Sample Preparation Area: The area is used for the extraction of nucleic acids. PCR products of DNA clones containing the sequence to be amplified should not be handled in this area. The following precautions should be taken while working in the sample preparation area:

1. Samples should be manipulated with specialized barrier or positive displacement pipettes which prevent the carry over of aerosols created during the pipetting.
2. Aerosols should be minimized by briefly centrifuging the tubes prior to opening.
3. Popping and tapping of the tubes during opening should be avoided.
4. For purification of template, simpler methods giving better results are preferable because of lesser manipulation of the samples.
5. Lab coats and gloves should be worn at all the times and gloves should be changed frequently.

During sample preparation in the RNA PCR, the following more precautions should be taken. Since there are more number of steps in the preparation, there are more chances of sample to sample contamination. Therefore, special precautions should be taken.

1. Use of Uracil N-glycosylase (UNG) is done with RNA PCR.
2. Reverse transcriptase step is carried out in the sample preparation area.
3. Reverse transcriptase step should be terminated by boiling.
4. Thermostable reverse transcriptase (rTth) may preferentially be used than the conventional two steps reverse transcriptase and Taq DNA polymerase. The advantages in using rTth are significant: There is a single buffer system; dUTP and UNG may be incorporated in the reverse transcriptase reaction; There is improved specificity of the reaction resulting from less mispriming because the cDNA synthesis is performed at a higher temperature with rTth. For laboratories handling large number of RNA samples for PCR, rTth system is better due to reduced handling and improved contamination control.

(b) **Pre-PCR Area:** This area is used to prepare individual reactions for PCR. The area should be clean and free from all the sources of possible contaminations in the process of molecular cloning and sample preparation. The primers and probes used in the PCR should not be prepared in this area.

(c) **PCR Area:** In this area, PCR machine with uninterrupted power supply should be installed. This area should also be neat and clean and free from any type of contamination.

(d) **Post-PCR Area:** It is the area separated from the PCR area where post reaction manipulations of sample, analysis and data interpretation is done. No equipment or reagent from this area of the laboratory should be shifted to any pre-PCR area. In PCR, main source of contamination is through micro-aerosols.

It is important to mention that in the PCR laboratory starting from the sample preparation area upto post PCR area, traffic should be unidirectional.

The following general but important precautions should be followed in a PCR laboratory:

1. All the plastic wares to be used should be of good quality, pre-sterilized and disposable.
2. All the solutions to be used should be free from contaminating nucleic acids and/or nucleases (both DNases and RNases).
3. High quality reagents free from heavy metals and nucleases should be used in PCR.
4. Water used in the preparation of the reagents and master mix etc should be distilled, filtered through 0.22 micron filter and properly autoclaved.
5. All the reagents should be made in large volumes tested for the performance and aliquoted into single usage volumes for storage.
6. Pipettes used in the sample preparation should be stored air tight when not in use.
7. Lab coats from the Pre-PCR area should not be taken anywhere else.
8. From a lab management perspective, if a fewer number of samples are to be tested, worker should do pre- PCR first than post-PCR.

Actual PCR Reaction

It requires the following steps:

1. **Initial denaturation:** A double stranded DNA should get completely denatured for complete unwinding of the DNA. It is done by heating at 90-95°C.
2. **Annealing of the Primer:** An oligonucleotide is used as a primer which gets annealed at the complementary site. The denaturation and annealing steps following the initial set up rarely need to be more than 60 seconds in duration. A temperature of 90-95°C is used for denaturation and 50°C for reannealing. It is true that higher the annealing temperature, the greater is the specificity of hybridization of the oligonucleotide primer.
3. **Extension of the Primer:** Extension of the primer is done at 72°C using all the four deoxyribonucleoside triphosphates, magnesium ions and heat stable DNA

polymerase. Final extension of the chain is done at 72°C for 5 to 10 minutes (Figure 9.1). Adding too much of DNA polymerase is avoided since it may increase non-specific amplification due to sometimes non-specific binding of the DNA primer in the presence of Taq DNA polymerase. The exact reason for non-specific binding is not known.

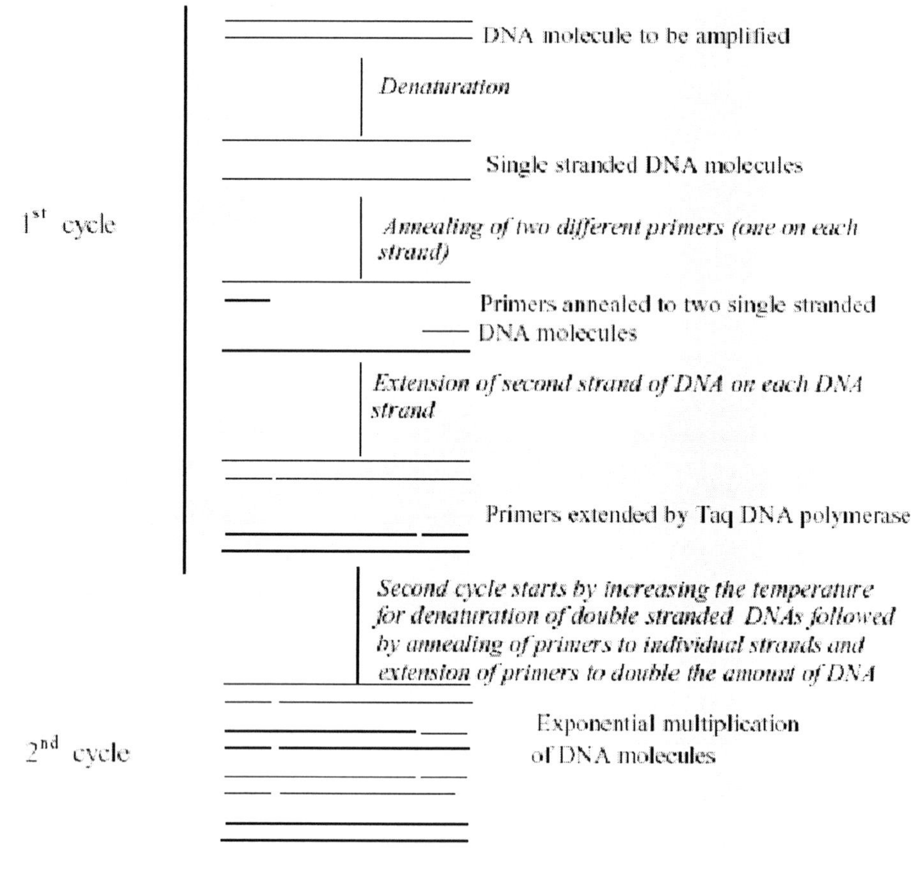

Figure 9.1. Schematic representation of PCR

Generally, for amplification of a sequence, between thirty to forty cycles are run in PCR machine.

The PCR reactions should always be carried out under a layer of sterile mineral oil to prevent evaporation of the reaction solution. As far as possible, the reaction volume is kept less than 100 microlitres to ensure rapid temperature equilibration. If a dry heat block is used, a little mineral oil should also be added to the wells to ensure good thermal contact. Samples of the reaction mixture may be withdrawn during the course of a PCR experiment in order to monitor the progress of the amplification.

4. **Visualizing results:** For confirming amplification of the DNA, 1-2% agarose gels or 5-10% polyacrylamide gels are used depending on the size of the DNA fragment

being amplified. The amplification of the correct DNA sequences carried out is tested either by doing appropriate restriction endonuclease digestions or by southern blotting followed by hybridization with a suitable probe. Southern blotting may reveal the amplified sequences which are not visible by ethidium bromide staining. Alternately, nested PCR or enzyme linked immunosorbent assay (ELISA) for confirmation of the product of PCR is performed.

A standard PCR reaction in 50 microlitres volume generally has the following:

- 5x Buffer (67 mM Tris-HCl, pH 8.8, 16.6 mM ammonium sulfate, 170 microgram/ml
- bovine serum albumin, 6.7 micro molar EDTA)
- 6.7 mM $MgCl_2$
- 10 mM 2-mercaptoethanol (optional)
- 200 microgram/ml dNTPs
- 10 nanogram linearized DNA
- 0.3 microgram primer
- Sterile distilled water to make up the volume
- One should use sterile, siliconized 0.5 ml capped microfuge tubes.

Researchers generally make a master mix tube while setting up a large number of PCR reactions since dispensing individual components in each tube is a time consuming, laborious and uneconomical. It requires many pipette tips, tubes and other supplies. Repeated pipetting also adds chance of error and contamination. Considering all these aspects, there are many software packages available that design PCR primers and optimize reaction conditions. Kumar and his associates developed a software for calculating different components of PCR reaction to be added to the master mix and then dispensed into the individual tubes and called it as PCR Recipe.

9.3. Primer Designing

The PCR follows relatively simple, predictable and well understood mathematical principles. Efficacy and sensitivity of PCR largely depend on the efficiency of the primers. The ability for an oligonucleotide to serve as a primer depends on several factors such as kinetics of association and dissociation of primer-template duplexes at the annealing and extension temperatures, duplex stability of mismatched nucleotides and their location. The primers which are unique for the target sequence to be amplified should fulfill certain criteria such as primer length, GC %, annealing and melting temperatures, 5' end stability, 3' end specificity etc.

Length of the primer is a critical parameter. Length of the primer should be such that ensures a denaturation temperature of 55-56°C. This enhances specificity and efficiency. For general studies, primers of typically 17-34 nucleotides in length are the best. Primer of more than 16 mers in length are not generally annealed to non-target sequences.

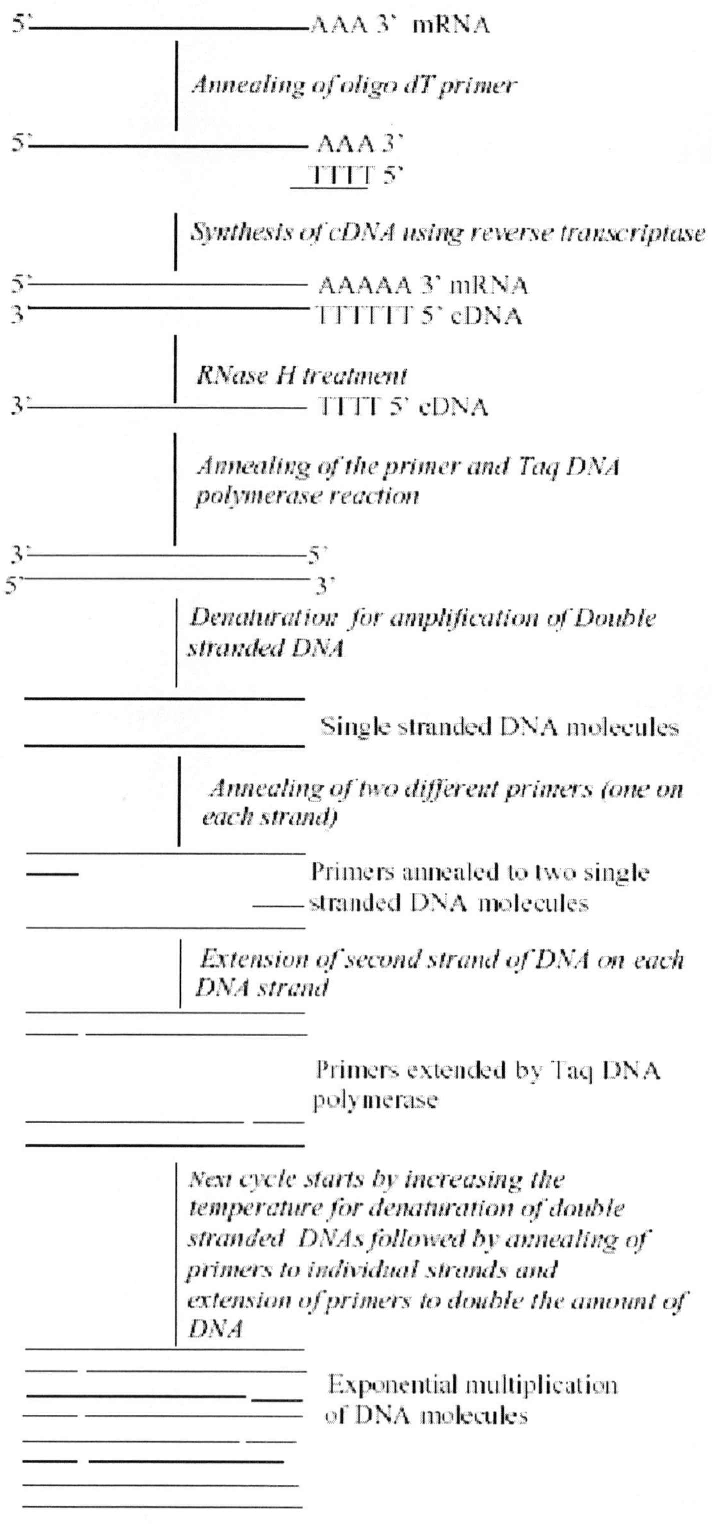

Figure 9.2. Schematic representation of RT-PCR

Both the terminals of the primer are of vital importance for a successful amplification. The 3' end position in the primer affects mispriming. Runs of C's or G's at the 3' end of the primer should be avoided as GC rich sequence leads to mispriming. Complementarity at the 3' end of the primer elevates mispriming as this promotes the formation of a primer dimer artifact and reduces the yield of the desired product. The stability of a primer is determined by its false priming efficiency, ideally it should have a stable 5' end and an unstable 3' end. If the primer has a stable 3' end, it will bind to a site which is complementary to the sequence rather than the target site and may lead to secondary bands. If G or C in last 3 bases at the 5' termini is present, it is good for efficient binding of the primer to the target site. GC clamp reduces spurious secondary bands.

GC content, melting temperature and annealing temperature are strictly dependent on one another. GC % is an important characteristic of DNA and gives information about the strength of the annealing. A GC of 50-60% is recommended.

An important factor to consider when designing a primer is the presence of secondary structures in the primer which reduces the number of the primer molecules available for bonding. The presence of hairpin loops reduces the efficiency by limiting the ability to bind to the target site. It is established that under particular conditions, the relative stability of a DNA duplex depends on its nucleotide sequence and more particularly on the identity of the nearest neighbour nucleotides. The overall stability and the melting temperature of a DNA duplex is predicted from its nucleotide sequence. Recently, it has been shown experimentally that hairpin loops, if present, can reduce the efficiency of the reaction by limiting primer availability.and the ability to bind to the target site.

Annealing between the 3' end of one primer molecule and the 5' end of another primer molecule and subsequent extension results in a background product known as primer dimer. Its subsequent amplification product can compete with the amplification of the larger target. If the primer binds anywhere else than the target site, the amplification efficiency gets reduced. To minimize dimers and false priming, PCR is generally performed at high temperature. Generally, all the reagents except one (Taq DNA polymerase) are mixed at the room temperature (master mix). The sample is denatured completely for 3 to 7 minutes, put on ice for a couple of minutes and afterwards DNA polymerase is mixed to start the reaction.

PCR product length is directly proportional to inefficiency of amplification. Primers should be designed so that only 150 to 1000 bp can be amplified which is ideal for probe hybridization studies. For RT-PCR, primers should be designed in exons only taking care that both primers should be on different exons of mRNA to avoid spurious product.

There are good and bad aspects of mismatches in primers. Single mismatch at the 3' end reduces efficiency of the amplification. However, on the other hand, mismatch 3-4 bases upstream of the 3' end of a primer used for amplification refractory mutation system (ARMS) study actually increases specificity. A mismatch may therefore be created while designing a primer for ARMS PCR. Nowadays, a number of softwares are available commercially which are much useful in designing desired primers. Universal random primers which are generally used for studying biodiversity using random amplification polymorphic DNA (RAPD) analysis have been designed. The primer(s) should be designed carefully for a successful PCR study.

9.4. Types of PCR

Competitive and Quantitative RT-PCR

Competitive RT-PCR - In this type of PCR there is no real competition in the biochemical sense as all reagents are in excess. The only limiting factor is the copies of cDNA taken in the first few cycles.

Quantitative RT-PCR - For quantifying mRNA, internal standard RNAs are added in a defined quantity to the RNA sample prior to the RT reaction. The resulting standard cDNA is co -amplified with the same primers as the endogenous target sequence. Its PCR product is approximately 50 nucleotides smaller. This method allows measurement of small differences in mRNA amount in the RNA samples.

(i) Reverse transcriptase PCR (RT-PCR): Although, PCR amplification is performed generally on the DNA template but RNA may also be used for amplification by using RT-PCR. The RT-PCR is useful for studying the expression of genes and for monitoring the obscure species of mRNA. It involves the use of random primer (usually hexamer) or oligo dT complementary sequence at the 3' end of the mRNA (poly A tail). If oligo dT primer is used, since it is complementary to the 3' poly A tail of the mRNA, it anneals to the mRNA used as the target. This form of PCR enables to quantitate mRNA levels and to generate cDNA fragments. The RNA strand is subjected to reverse transcription by the enzyme reverse transcriptase in a PCR reaction. The cDNA synthesized can then be used for PCR amplification using the conventional PCR technique. After annealing, the RNA strand is reverse transcribed using reverse transcriptase enzyme. As a result, cDNA is synthesized which is further used in PCR amplification (Figure 9.2). Therefore, in RT-PCR, there are two steps, reverse transcriptase and Taq DNA polymerase. Nowadays, people have taken advantage of the property of reverse transcriptase to use RNA as well as DNA as template. Using thermostable reverse transcriptase, RT-PCR may be carried out as one step process.

RT PCR is particularly useful for studying the expression of genes and for monitoring obscure species of mRNA. It is also used for cloning or subcloning and for the identification of bacterial colonies.

(ii) Inverse PCR: Conventional PCR allows the amplification of a DNA segment flanked by the two primers but this does not permit the acquisition of sequence information outside the boundary defined by the primers. This limitation is circumvented by using inverted or inverse PCR. In inverse PCR, instead of two primers facing each other as in the regular PCR, the two primers face outwards. The target DNA is digested with a restriction endonuclease or sheared by physical means to generate the linear DNA. The linearized DNA is then circularized by DNA ligase through an intramolecular ligation reaction. The two primers annealed with the circular DNA now face each other flanking the unknown sequences. Therefore, using inverse PCR, the unknown sequences flanking a known sequence can be readily amplified.

(iii) Nested PCR: It involves one PCR reaction followed by the next PCR extension which amplifies the first PCR product. Two sets of primers are designed. PCR is first carried out using outer primers and subsequently with inner primers positioned with in the product obtained in the first extension. While designing primers for nested PCR, care should be taken to eliminate potential primer dimers and cross dimers with in and between inner and outer primer sets. Nested PCR is successful in reducing unwanted products and increases sensitivity. It is generally used when the actual quantity of target DNA is very low or when the target DNA is impure.

(iv) AP-PCR: Arbitarily primed polymerase chain reaction (AP-PCR) requires only a single primer of relatively much smaller length as compared to the primers used in the conventional PCR. The single primer anneals to both the strands arbitrarily and fragments of various lengths are randomly generated giving rise to smeary pattern when the product is resolved on the agarose gel. The primer length, sequence complexity and the available annealing sites in the target DNA affect the overall amplicons. The AP-PCR approach has been used as a powerful adjunct in the context of DNA profiling and found to be useful in animal and plant biotechnology as well as in forensic medicine.

(v) Multiplex PCR: It involves co-amplification of two or more target sequences within a single sample. A unique pair of primers for each target is preferred but primers can be designed so that a single primer can amplify different regions with two or more counterparts. While designing primers for multiplex PCR, the basic rule is to have similar annealing temperatures and similar GC% of the primers. Multiplex PCR may be used for detection of genetic disorders.

(vi) ARMS –PCR: ARMS is also known as allele-specific PCR (ASPCR) or PCR amplification of specific alleles (PASA). It was first described by Newton and co-workers in 1989. ARMS PCR is used to detect point mutations and small deletions/insertions in the genomic DNA. It may also be used in HLA (human leukocyte antigen) typing. ARMS PCR can generally detect a single copy of the mutant allele amongst 40 copies of the normal. It generates rapid, reproducible and inexpensive results.

Principle of the technique. A PCR primer is designed so that it can discriminate between templates which differ at a specific single nucleotide residue. It is based on the principle that Taq polymerase has no 3' to 5' exonuclease activity; therefore a mismatch at the 3' end of the primer and the template will result in its inability to function as a primer under appropriate conditions.

A standard ARMS PCR consists of two complementary reactions (two tubes) and utilizes 3 primers. One primer is constant and complementary to the template in both reactions, the other primers differ at their 3' terminal residues and are specific to either the wild type DNA sequence or the mutated sequence at a given base. Only one of these primers is used per tube. If the sample is of a homozygous mutant or homozygous wild type, amplification will only occur in only one of the tubes, if the sample is heterozygous, amplification will be seen in both the tubes.

The ARMS primer is designed so that the nucleotide distinguishing between the different alleles is at the 3' terminal base of the primer. Ideally a control set of primers should be included in the PCR as an internal amplification control. ARMS Primers are usually ~30 bases long (14-30 have been used). Low dNTPs concentrations result in a more specific ARMS reaction (100μM). Reducing oligonucleotide (primer) concentrations to 0.05 μM may increase specificity.

Modifications of ARMS Methodology

Multiplex ARMS PCR. In this method a number of ARMS primers, each specific for a particular allele, are included in a single reaction. PCR products corresponding to different alleles are separated due to physical characteristics viz. length. This modification exempts the need for control primers to be added to the reaction and ARMS reactions act as internal controls for each other. Besides this allows a bi-allelic system to be genotyped in one tube (4 primers: 2 ARMS primers- allele specific and 2 Non ARMS primers and a control product is obtained from the 2 non ARMS primers).

Double ARMS PCR. Two allele specific primers are used simultaneously in a single reaction (no Non-ARMS primers used in the reaction). Amplification only occurs when the alleles specific for each of the ARMS primers are present on the same chromosome. Double ARMS PCR is a powerful tool in clarifying haplotypes, particularly in individuals who are doubly heterozygous at two polymorphic sites. Double ARMS PCR is useful in human leukocyte antigen (HLA) typing. Double ARMS PCR has been shown to detect a single target molecule in an excess of 10^5 background molecules. Hence, this technique is used to detect fetal cells in maternal blood.

Fluorescent ARMS PCR. The wild type and mutant primers are labeled with different fluorochromes, and are then incorporated into the same reaction tube. This is a more robust test, having much increased specificity.

(vii) Asymmetric PCR: In this modified approach, target sequence of one strand can be amplified several orders of magnitude as compared to its complementary strand. In this type of PCR the quantity of one of the primer corresponding to one strand is reduced hundred folds. The primer for the other strand of interest is present in normal quantity.

As PCR starts, the initial amplification cycles of both the strands proceed at the same speed and amplicons are generated from both the strands. However, as the reaction proceeds and the number of DNA strands to be copied increases, the strand which has a reduced quantity of primers lags behind the other since the number of primer molecules present for binding to the large number of newly generated fragments is lesser. On the other hand, the amplicons from the other strand are generated abundantly. Thus when the PCR products are observed after electrophoresis, the bands obtained are as the result of the amplification of regions of one strand only as the very small number of amplicons generated from the other strand cannot be visualized. This approach is used for generation of single stranded DNA fragments in large amounts which are to be used for many purposes viz. DNA sequencing.

Real time PCR: In PCR, as little as a single copy of DNA fragment can be specifically amplified and detected. Theoritically, there is a quantitative relationship between amount of starting target sequence and amount of PCR product at any given cycle. However, in practice, different amounts of PCR products are obtained. The development of real time quantitative PCR allows the routine and reliable quantification of PCR products. This provides the opportunity to perform sensitive, accurate and reproductive measurements of levels of gene expression. Real time PCR may also be used to measure viral load, to perform allelic discrimination studies etc. In Real time systems for PCR, use of probe rather than ethidium bromide intercalator dye has been taken for detection of PCR products. The main drawback in intercalator dye based detection of PCR product is that both specific and non-specific products generate signal. An another method, the 5' nuclease assay provides a real time method for detecting only specific amplification products. During amplification, annealing of the probe to its target sequence generates a substrate that is cleaved by the 5' nuclease activity of Taq DNA polymerase when the enzyme extends from an upstream primer into the region of the probe. This ensures that cleavage of the probe occurs only if the target sequence is being amplified.

Nowadays, development of fluorogenic probes made it possible to eliminate post-PCR processing for the analysis of probe degradation. The reporter fluorescent dye and a quencher dye are attached on the oligonucleotide. In intact probe, the proximity of the quencher greatly reduces the fluorescence emitted by the reporter dye by Forster resonance energy transfer through space. The reporter dye is tagged at the 5' end and the quencher dye at the 3' end of the primer. The probe anneals downstream from one of the primer sites and is cleaved by the 5' nuclease activity of Taq DNA polymerase as this primer is extended. This cleavage separates the fluorescent dye from the quencher dye increasing fluorescent signal. Cleavage removes the probe from the target strand whereas primer extension is continued. The inclusion of the probe does not inhibit the overall PCR process. More reporter dye molecules are cleaved from their respective probes with each cycle, increasing fluorescence intensity proportional to the amount of amplicon produced. Since specific hybridization between the probe and target is required to generate fluorescent signal, non-specific amplification due to mis-priming or primer-dimer artifact does not generate signal. Besides, probes can be labeled with different reporter fluorescent dyes, then amplification of two distinct sequences can be detected in a single PCR reaction. The disadvantage of fluorogenic probes is that different probes must be synthesized to detect different sequences. Nowadays, specific equipments are available for real time PCR which have built in thermocycler and fluorescence detection system.

9.5. Applications of PCR

PCR has a wide range of applications and only some of these are listed here:

1. Amplification of human specific DNA sequences
2. Differentiation of species, sub species and strains
3. PCR amplification for DNA sequencing
4. For detection of mutations
5. For monitoring cancer therapy
6. To detect bacterial and viral infections
7. For pre-determining the sex
8. For linkage analysis using single sperm cells
9. For studying molecular evolution
10. For ascertaining the recombinant clones

9.6. Limitations of PCR

PCR is a sensitive technique and therefore is highly susceptible to contamination which can result in false positivity. Various measures are recommended to reduce contamination problems such as single use of the reagents, different rooms for DNA specimen preparation, PCR amplification and post-PCR detection, use of UV or gamma radiation to inactivate or destroy contaminating DNA. Over and above, a clean work habit ensures trouble free and reliable results.

9.7. Precautions of PCR

The PCR is capable of amplifying as little as a single molecule of DNA, therefore, precautions should be taken to guard against contamination of the reaction mixture with trace amounts of DNA that could serve as templates. Such precautions are essential when amplifying target sequences are present at low concentrations. Some important precautions to be followed are listed here:

1. If possible, assemble and carry out PCR in a laminar flow hood equipped with ultraviolet lights. Turn on these lights whenever the hood is not in use. Keep a microfuge, disposable gloves, supplies and sets of pipetting devices used only to handle PCR in the hood. Since the barrels of automatic pipetting devices are common sources of contamination, prepare and handle reagents using positive displacement pipettes equipped with disposable tips and plungers. Autoclave all buffers, pipette tips and centrifuge tubes before use.
2. Put on a fresh pair of gloves when starting work in the area designated for assembly of polymerase chain reactions and change gloves frequently.
3. Prepare your own set of reagents and store them in small aliquots preferably in a specially designated section of a freezer located near the hood. Do not use these reagents for other purposes. When preparing these reagents, use new glasswares, plastiwares and pipettes that have not been exposed to DNAs in the laboratory. After use, discard aliquots of the reagents. These should not be returned to storage vials.
4. The microfuge tubes containing reagents to be used in polymerase chain reactions should be centrifuged briefly (say for 10 seconds) in a microfuge located in the laminar flow hood before opening them, The centrifugation deposits all the fluid in the base of the tube and reduces the possibility of contamination of gloves or pipetting devices.
5. It is better to add all the components of the reaction including the mineral oil used to prevent evaporation to the microfuge tube before adding the template DNA. Afterwards, template DNA should be added and the tube should be closed immediately. The contents should be mixed by gently tapping the side with a gloved finger. The tube should be centrifuged briefly for 10 seconds to separate the aqueous and organic phases.

6. While adding the template DNA to an assembled polymerase chain reaction, take care not to create aerosols that could contaminate other reactions. Keep the lids tightly closed on all the tubes that are not in immediate use. Change gloves after handling tubes containing template DNA.
7. Whenever possible, use a positive control (a polymerase chain reaction that contains a small amount of the appropriate target sequence). An appropriate dilution of the target sequence should be prepared ahead of time in a different area of the laboratory. This prevents the introduction of concentrated solution of target DNA into the area of the laboratory set aside for polymerase chain reactions.
8. Always include a control that contains all the components of the polymerase chain reaction except the template DNA. Assemble this control after all other polymerase chain reactions have been set up.

9.8. Modifications of PCR

i. Ligase chain reaction (LCR): It requires a thermostable ligase and allows the discrimination of DNA sequences differing in only single base pair (single point mutation). Therefore, it has become an important diagnostic tool that can be used in conjunction with a primary PCR amplification. The principle of LCR is based on the ligation of two adjacent oligonucleotide primers that uniquely hybridize to one strand of the target DNA. The two primers are usually positioned such that the nucleotide at the 3' end of the upstream primer coincides with the targeted single base pair difference. This single base pair difference may define two different alleles, species or other polymorphisms correlated to a given phenotype. If the target nucleotide at that site complements the nucleotide at the 3' end of the upstream primer, the two adjoining primers can be covalently joined by the ligase. The important feature of LCR is a second pair of primers having almost complementarity to the first pair and are designed with the nucleotide at the 3' end of the upstream primer denoting the sequence difference. Using ligase, the two primers can be ligated and the ligated products can serve as templates for the next reaction cycle leading to an exponential amplification process. If there is a mismatch at the primer junction, it will be discriminated against by thermostable ligase and the primers will not ligate. The absence of the ligated product will indicate a single base pair change in the target sequence.

LCR assays have been developed for the detection of genetic diseases as well as bacterial and viral diseases. Many times, LCR is preceded by an initial PCR step to get more sensitivity of the assays. PCR products may be used in LCR without prior purification. Due to the ability of LCR for detection of point mutation, its importance increased from the time of availability of the genome sequences from human and other species. LCR may be useful for the diagnosis of pathogenic bacteria and viruses in many ways. It may be used for screening antibiotic resistance arising from the point mutations. In viruses, the identification of subpopulations with genetic differences may be important with respect to host range, drug resistance and virulence.

ii. Strand displacement amplification (SDA): It is the technique used for the isothermal amplification of DNA targets. A restriction endonuclease and a strand displacing DNA polymerase, both capable of operating at 37°C are used. In this technique, there is continuous primer extension of DNA targets with displacement of the nascent DNA strands by the DNA polymerase and subsequent capture of these strands by new primer. The restriction enzyme produces a single stranded nick on the double stranded strand displacers. The nick serves as a new primer for the next round of primer extension by DNA polymerase. During next round synthesis, the downstream DNA strand is displaced concomitant with the synthesis of a new strand from the nick. As a result, the reaction products from the sense strands feed into the antisense strand and *vice versa*.

It requires four primers, two enzymes, a dNTPs mixture where one dNTP is substituted by a phosphorothioate dNTP and other reaction components. Target DNA is briefly denatured at 95°C and subsequently cooled to 37°C, afterwards, the two enzymes are added and two hour incubation at 37°C amplifies the target DNA to about 10^7 fold amplification. The amplified DNA may be detected using any conventional method.

The dual enzyme system in SDA allows amplification to occur under isothermal conditions. It is a simple and low cost technique compared to PCR which requires thermocycling. Thermocycling based amplification systems are sensitive to both the absolute temperature and time, therefore, variation from well to well in the thermocycler can affect uniformity of the method.

The technique requires DNA polymerase that incorporates dNTPs with efficiency and lacks 5',3' exonuclease activity. It should be able to start DNA synthesis from the nicked restriction site and should displace nascent DNA strands in a 5',3' direction. Therefore, commercially available Klenow fragment works well. Large fragment of *Bacillus stearothermophilus* DNA polymerase also works satisfactorily. Many restriction endonucleases may generate a single stranded nick in their hemiphosphorothioate substituted recognition sites but only few work in SDA since recognition site must be non-palindromic.

iii. Transcription mediated amplification (TMA): Most of the amplification methods involve DNA replication. The alternative systems based upon RNA intermediates have been developed. Several of the methods exploit the ability of phage DNA directed RNA polymerases to reinitiate multiple times upon a double stranded template containing the cognate phage promoter. These RNA based methods also require interconversion of the transcripts to DNA through reverse transcriptase. These methods require at least one amplification primer to contain a phage promoter 5' to the target binding region of the primer. There is isothermal amplification. The system utilizes RNA targets, primers containing T7 RNA polymerase promoter sequence, reverse transcriptase and a reaction cocktail of dNTPs and NTPs. In the first round, primers bound to the RNA target are incubated at 42°C with reverse transcriptase to produce cDNA from each target molecule at 42°C. Following heating at 100°C and equilibration at 42°C, more reverse transcriptase is added which synthesizes double stranded cDNA. This cDNA acts as substrate for RNA polymerase that is added next. By the process, multiple transcripts of each cDNA may be produced. The procedure has been used to amplify HIV nucleic acid.

In an improvement system, RNase H was also added which degrades RNA component of RNA-DNA hybrid. This resulted in more complete synthesis of cDNA (Guatelli et al., 1990). The method has been used as diagnostic method for detection of HIV. The system has been further improved by using 10% DMSO and 15% sorbitol in the reaction and under the conditions, only two enzymes have to be used, RNA polymerase and reverse transcriptase. DMSO and sorbitol stimulate the endogenous RNase H activity of reverse transcriptase. It has also been shown that even lesser amount of these two enzymes are needed in the system.

References

[1] Dieffenbach, C.W. and Dveksler, G.S. (1995) *PCR Primer: A laboratory Manual*, Cold Spring Harbor Laboratory Press, New York.

[2] Singh, V.K. and Kumar, A. (2001) PCR Primer Design. *Mol. Biol. Today* 2: 27-32.

[3] Singh, V.K., Govindrajan, R., Naik,S. and Kumar, A. (2000) The effect of hairpin structure on PCR amplification efficiency. *Mol. Biol. Today* 1: 67-69.

[4] Singh, V.K. and Kumar, A. (2000) PCR Recipe: Software for setting up PCR reactions. *Biotech Software & Internet Report.* 1: 276-277.

[5] He, Q., Marjamaki, M., Soini, H., Mertsola, J. and Viljanen, M.K. (1994) Primers are decisive for sensitivity of PCR. *Biotechniques.* 17: 82-87.

[6] Wu, D.Y.W., Ugozzoli, B.K., Pal, J., Qian, and Wallace, R.B. (1991) The effect of temperature and oligonucleotide primer length on specificity and efficiency of amplification by the polymerase chain reaction. *DNA Cell Biol.* 10: 233-238.

[7] Huang, M., Arnheim, N. and Goodman, M.F. (1992) Extension of base mispairs by Taq DNA polymerase: Implications for single nucleotide discrimination in PCR. *Nucleic Acid Res.* 20: 4567-4573.

[8] Guatelli, J.C. Whitefield, K.M., Kwoh, D.Y., Barringer, K.J., Richman, D.D. and Gingeras, T.R. (1990) Isothermal in vitro amplification of nucleic acids by a multienzyme reaction modeled after retroviral replication. *Proc. Natl. Sci. Acad. USA.* 87: 1874-1878.

Chapter X

Gene Therapy

10.1. Introduction

Gene therapy is the process by which DNA sequences encoding specific genes are delivered to cells with the aim of treating or curing a disease. The concept of gene therapy developed within the perspective of "molecular medicine" as it became evident that many diseases arise because of an individual is deficient for a critical protein that is encoded by a particular gene. Since each protein is produced according to instructions encoded in the corresponding gene, such diseases arising from mutations in genes were designated as "genetic diseases." Classical examples of such diseases (with a genetic origin) are forms of diabetes due to lack of insulin (a protein hormone), Duchenne muscular dystrophy, hemophilia due to the lack of factor VIII or factor IX (proteins required for blood clotting). Factors VIII and IX are synthesized in the liver and participate in the chain of events that produces a long fibrous protein called fibrin that serves as the matrix for blood clot formation. In hemophilia, Minor cuts and bruises, which would do little harm to most people, can prove fatal to hemophiliacs, who lack the proteins (Factors VIII and IX) involved in the clotting of blood due to deletion or presence of defective genes. Sadly, before these proteins were made available through genetic engineering, hemophiliacs were treated with proteins isolated from human blood. Some of this blood was contaminated with the AIDS virus, and has resulted in tragic consequences for many hemophiliacs. Now, use of genetically engineered proteins in therapeutic applications is common rather than blood products. Often a mutation occurs in gene that eliminates the formation of such a protein in a functional form resulting in a disease state. Therefore, gene therapy concerns the possibility of delivering a normal gene to the tissues or taking corrective measures to where a mutant gene has rendered the corresponding protein non-functional. If the gene therapy is successfully achieved, the individual suffering from a genetic disease would have the deficiency corrected and lead a normal life. Recent advances in genomics and molecular biology have revealed that almost all diseases have a genetic component. In cystic fibrosis or hemophilia, mutations in a single gene result in disease. Whereas in hypertension or high cholesterolemia, certain genetic variations may interact with environment stimuli to cause disease. Pathological conditions associated with ageing frequency result from the loss of gene activity in specific types of cells.

Some single gene disorders are quite common-cystic fibrosis is found in one out of every 2,500 babies born in the Western World-and in total, diseases that can be traced to single gene defects account for about 5% of all admissions to children's hospitals. It is expected tat the genetic disease may be corrected with the use of gene therapy.

The first example of gene therapy in a mammal was the correction of the growth hormone deficiency in mice. The recessive mutation *little* (lit) results in dwarf mice, even though the mouse growth hormone gene is present and apparently normal, no messenger RNA (mRNA) is produced. The initial step in correcting this deficiency was to inject homozygous *lit that is lit* eggs with about 5000 copies of a 5-kb linear DNA fragment that contained the rat growth hormone structural gene (RGH) fused to a regulator promoter sequence from a mouse metaloprotein gene (MP). The eggs were then implanted into pseudopregnant mice, and the baby mice were raised. About 1% of these babies turned out to be transgenic, showing increased size when heavy metals were administered during development. A representative transgenic mouse was then crossed to a *it lit* female. The site of insertion of the introduced DNA in mammals is highly variable, the DNA is generally not found at the homologous locus.

10.2. Diseases of Genetic Origin

The complete human genome sequence has already been completed. This work will enable scientists and clinicians to understand the genes that control all diseases to which the human race is prone, and hopefully develop new therapies to treat and credit diseases. If the gene is dominant, it alone can produce the disease in the couples even if its counterpart is normal, thus the children of a parent with the disease can be affected. Huntington's chorea, a severe disease of the nervous system, which becomes apparent only in adulthood, is an example of a dominant genetic disease. A potential approach to the treatment of genetic disorders in man is gene therapy. This is a technique whereby the absent or faulty gene is replaced by a working gene, so that the body can make the correct enzyme or protein and consequently eliminate the root cause of the disease.

The most likely candidates for future gene therapy trials will be rare diseases such as Lesch-Nyhan syndrome a distressing disease in which the patients are unable to manufacture hypoxanthine-guanine phosphoribosyl transferase enzyme. This leads to a bizarre impulse for self-mutilation, including very severe biting of the lips and fingers. The normal version of the defective gene in this disease has now been cloned.

10.3. Types of Gene Therapy

Gene Augmentation Therapy

The traditional approach to gene therapy is to introduce DNA to the genome to replace a defective DNA/gene fragment that has lost normal function, and can be described as gene augmentation therapy (GAT). The aim of gene augmentation therapy is to correct a loss of function caused by a deletion by adding the functional allele. Transferred genes may be stably integrated in to the genome with the potential to permanently correct the defect, especially if the stem cells transformed may be maintained episomally (an inevitable decay in the maintenance of gene expression and treatment may need to be repeated). Several GAT clinical trials are currently underway for diseases including cystic fibrosis, adenosine deaminase deficiency and familial hypercholcstrolcmia.

Gene Inhibition Therapy at the Nucleic Acid Level

For the treatment of dominant genetic disease mutations causing loss of gene function or mutations that lead to some gain of function, gene augmentation therapy is less powerful. In principle, a valid approach to the treatment of such diseases would be targeted correction (i.e. allele replacement or gene knockout to remove the mutant allele). However, this is a very inefficient process in practice, even in cultured cells, and its application to the correction of genetic defects in many somatic cells, especially *in vivo*, demands further technical improvements. A novel approach, in future, is targeted correction at the RNA level by using ribozymes or RNA editing enzymes to correct pathogenic mRNAs.

An alternative strategy, which is presently undergoing clinical trials for the treatment of several types of cancer, is the use of nucleic acids to inhibit gene expression. The advantage of this approach is that the inhibitor can be synthesized or developed to inhibit the specific allele, without effecting the expression of any normal functioning allele. The introduction of antisense genes allows the stable and permanent expression of antisense RNA, which binds to (mutant) mRNA and prevents translation (it may also target the mutant mRNA for degradation). Furthermore, increasing use is being made of antisense construct containing ribozymes, which degrade the mRNAs to which they bind. Oligonucleotides can be used for epigenetic gene therapy (i.e. therapy which does not involve changes to the genome) and can act in two ways. Antisense oligonucleotides act in the same way as antisense RNA, by binding to mRNA and causing inhibition and degradation (in this case, probably by recruiting RNAs which digest the RNA strand of DNA: RNA hybrids). Secondly, pyrimidine rich oligonucleotide can potentially form triple helix structures with purine rich strands of DNA (by Hoogsteen base pairing), which blocks transcription (triple helix therapy). Peptide Nucleic Acid (PNA) is an analog of DNA where the phosphate backbone is replaced by a neutral peptide backbone. PNA Oligonucleotides can be used as probes as well as agents for gene therapy. They form very stable triple-helix structures.

Gene Inhibition Therapy at the Protein Level

Targeted inhibition of gene expression can also take effect at the protein level. The cell is made to express within a cell of genetically engineered antibodies (intrabodies) which bind to and inactivate mutant proteins. Antibodies are not the only molecules being developed for protein level gene therapy. Any protein acting as multimer is a potential target for dominant negative inhibitor proteins, and this strategy is being explored to prevent viral coat protein assembly, including that of the HIV. A novel approach is to use degenerate oligonucleotides to identify specific oligonucleotide sequences, which interact with proteins. These Oligonucleotides or aptamers can then be used to inactivate specific mutant proteins.

Somatic Gene therapy focuses only on the body, or soma, attempting to effect a reversal of the disease phenotype by treating some somatic tissues in the affected individual. At present it is not possible to render entire body transgenic, so the method addresses disease whose phenotype is caused by genes that are expressed predominantly in one tissue. In such cases it is found that not all the cells of that tissue need to become transgenic; a portion of transgenic cells can ameliorate the overall disease symptoms. The methods proceed by removing some cells from an individual with the defective genotype and making these cells transgenic through the introduction of copies of the cloned wild type gene. The cells are then reintroduced into the patient's body, where they provide normal gene function. This is *ex vivo* gene therapy, gene transfer is performed in cells that have been removed from the patient. Once the therapeutic gene has been introduced into these cells they are transferred back to the patient.

For *in vivo* gene therapy approaches, genes are transferred into cells within the patient's body. The route of administration depends on the target tissue that needs to be treated as well as the mechanism through which the therapeutic gene elicits its effect. For example, gene therapy vectors for cystic fibrosis, a disease which effects cells within the lung and airway, are inhaled. Most gene therapies under evaluation to treat cancer are injected directly into tumors. Secreted proteins, such as Factor VIII or Factor IX for hemophilia or insulin for diabetes, may be introduced into the liver or muscle tissue.

Another approach the germinal gene therapy is the more ambitious to introduce transgenic cells into the germ line as well as into the somatic cell population. Such germinal therapy has been achieved using mice eggs. However, the protocol that is more relevant for application to humans is removal of an early embryo (blastocyst) with a defective genotype from a pregnant mouse and injection with transgenic cells containing the wild type allele. These cells become part of many tissues of the body, often including the germ line, which will give rise to the gonads. Then the gene can be passed on to some or all progeny, depending on the size of the clone of transgenic cells that lodges in the germinal area. However, no human germinal gene therapy has been performed to date.

Beyond simply replacing a gene, a further phase can also be considered that encompasses gene therapy, but is more precisely called cell therapy. In several hundreds of cases during the last decade, individuals with sickle cell anemia were "cured" for their lifetime (but they could still pass the mutated gene on to their children, although in this case without severe consequences if the other parent is normal) by bone marrow transplant. This procedure is similar to other forms of organ transplant in which the individual must also be treated with drugs that suppress immune responses in order to reduce the risk of rejection due to tissue incompatibility. In such procedures the donor is usually a close relative without the sickle mutation. In July 2000, a group in Edmonton, Canada reported treating diabetic patients by injecting normal pancreatic cells from brain-dead donors into the livers of the seven treated patients. These patients must also be maintained on drugs that reduce the risk of rejection, but have been able to live without insulin treatment for months following the operation.

10.4. Gene Delivery Systems

The problem of delivering a gene to a target cell is one of the principal obstacles that must be overcome for gene therapy. Since viruses are known that efficiently infect cells, the use of such viruses has generally been considered as the promising technique for delivering genes. In all cases, the virus is modified in such a way as to destroy its capacity to transmit disease or harm the cell and the gene to be delivered is incorporated into the genetic material of the virus. This new virus is called a "vector" since it can transfer the gene to the target cell. Once inside the cells of the individual treated (the host), the viral DNA recombines with host DNA and is thus integrated into the host chromosome.

The success of this approach has been limited for several reasons. First, the genes delivered by the viruses do not always work as well in the target cells as was hoped, since they are not necessarily used by the cell to produce sufficient amounts of the needed protein. Although the gene may well be inserted into the DNA of the target cells, not all position on the chromosome permit gene expression (i.e. transcription and translation into protein) with the same efficiency. Therefore, additional developments are required to target the genes transferred to portions of the chromosome that lead to more active expression. In addition, compared to blood cells, most other types of cells are difficult to isolate or target by the vector. Moreover, in certain cases, an immunological reaction is generated against the virus that limits its usefulness. Following the deaths of patients in recent trials in the United States it is now widely accepted that further studies on animal models are needed and that human subjects should only be considered under very favorable circumstances, such as were available to the group of Fischer in their work on SCID (severe combined immunodeficiency disease).

As the genetic and molecular basis for a multiplicity of diseases is elucidated, the promise of gene therapy continues to grow. Although initial efforts in gene therapy focused on delivering a normal copy of a missing or defective gene, current programs are applying gene delivery technology across a broader spectrum of disease conditions. Gene delivery is now being used to:

1. Replace missing or defective genes
2. Deliver genes that catalyze the destruction of cancer cells or cause cancer cells to revert back to normal tissue
3. Deliver viral or bacterial genes as a form of vaccination
4. Deliver genes that promote the growth of new tissue or stimulate regeneration of damaged tissue.

The application of gene delivery technology to a growing roster of clinical indications is predicted on significant advances in both genomics and gene delivery systems. As an increasing number of genes are identified and assigned to specific biochemical pathways, the therapeutic utility of these genes becomes more apparent. In parallel with an enhanced understanding of the roles that genes play in health and disease, improved vehicles, or vectors, for delivering therapeutic genes to target cells have been constructed. A growing number of genes are being evaluated for use as therapeutics. A few genes evaluated are given in the table below:

Gene	Indication	Function
CFTR	Cystic Fibrosis	Regulates intracellular chloride
Factor VIII	Hemophilia A	Regulates blood coagulation
Factor IX	Hemophilia B	Regulates blood coagulation
E1A	Cancer	Causes cancer cells to undergo cell death or revert to normal
P53	Cancer	Causes cancer cells to undergo cell death
AC6	Heart Failure	Increases contractile properties of the heart
VEGF	Coronary artery disease	Peripheral vascular disease induces the growth of new blood vessels
Interleukin-2	Cancer	Stimulates the immune system to kill cancer cells

Therapeutic genes can be introduced into target cells in a variety of ways. Gene delivery can be undertaken in cells that have been removed from the body (*ex vivo* gene therapy) or in the body itself (*in vivo* gene therapy). Genes can be introduced into cells using viruses, lipid-based (synthetic) vectors, naked DNA or electroporation (the application of an electrical charge, which creates opening in cell membranes). The selection of a particular delivery vehicle depends on a number of factors, including target cell type, duration of gene expression required for therapeutic effect, the size of DNA encoding the therapeutic gene and the route of administration. While no single vector developed to date is optimal for all clinical indications, the growing number of viral and synthetic vectors will enable gene therapy to be used in treating a wide variety of significant diseases.

10.5. Viral Vectors

Currently there are two ways of getting the transgene into the defective somatic cells. Both methods use viruses. The older method uses a disarmed retrovirus with the transgene spliced into its genome, replacing most of the viral genes. The natural cycle of retroviruses involves integration of the viral genome at some location in one of the host cells chromosomes. The recombinant retrovirus will carry the transgene along with it into the chromosome. This is a potential problem because the integrating virus can act as an insertional mutagen and inactivate some unknown resident gene, causing a mutation. This procedure has been used for somatic gene therapy of severe combined immunodefiency disease (SCID). In the gene therapy, blood stem cells were removed from the bone marrow, the transgene was added and the transgenic cells were reintroduced into the blood system.

Another vector used in human gene therapy is adenovirus. This virus normally attacks respiratory epithelia, injecting its genome into the epithelial cells. The viral genome does not integrate into a chromosome but persists extrachromosomally in the cells. This eliminates the problem of insertional mutagenesis by the vector. Another advantage of adenovirus as a vector is that it attacks nondividing cells, making most tissues susceptible in principle. Since cystic fibrosis is a disease of the respiratory epithelium, adenovirus is an appropriate choice of vector for treating this disease, and gene therapy for cystic fibrosis is currently being attempted using adenovirus. Viral vectors take advantage of a virus ability to enter cells and deliver genetic material to the nucleus. In developing viral vectors, DNA encoding some or all of the viral genes is removed and is replaced with a therapeutic gene. Most viral vectors also are engineered so that they are able to enter cells but lose their ability to replicate once inside. Based on the virus, from which it is derived, each type of viral vector has the ability to enter specific types of cells. While first generation viral vectors were thus restricted in their utility, newer vectors have been modified to meet specific objectives. In some cases this may mean engineering the vector so that it enters only one cell type while in other situation an expanded targeting capability may be desired. Modifications are made by changing some of the proteins that are expressed on the virus surface. These proteins interact with other molecules on the surface of target cells, similar to a lock-and-key combination.

In addition to varying targeting properties, different vectors have specific characteristics that regulate whether they permanently become part of the host cell's genetic material or exist transiently within the host cell. Depending on the therapeutic aim of a particular gene therapy, persistent or transient expression may be more or less desirable.

10.6. Synthetic Vectors

In addition to using viruses as vectors for gene delivery, a variety of non-viral vectors also are under development. These synthetic vectors generally are made up of lipids (fat molecules). Cell membranes contain a very high concentration of lipids. When a synthetic vector encounters the membrane of a cell, the lipids in the vector are incorporated into the lipids of the membrane, allowing the DNA contained within the vector to gain entry to the cell. This process is similar to how droplets of fat floating on water coalesce to form larger drops. While some synthetic vectors encompass a circle of DNA carrying a therapeutic gene, other systems also compact thc DNA. This compaction makes the vector particles smaller and of more uniform size. Just as a cell contains proteins and other molecules embedded in the lipids of its membrane, synthetic vectors can be engineered to carry targeting molecules on their surfaces. Such targeting moieties allow synthetic vectors to be directed to specific types of cells, which is important from the standpoint of both safety and systemic delivery.

10.7. Gene Expression Systems

In addition to choosing a relevant therapeutic gene and the appropriate gene delivery system, the ability to express (turn on) the delivered gene is a key factor in the development of successful gene therapies. Current approaches utilize a variety of promoters, DNA sequences that act as on/off switches for gene expression. Some promoters are active only in specific cell types, and are used to target gene activity to specific cells. Other promoters are expressed in a wide variety of cell types and may be used when cell-specific expression is not required. Promoter selection also depends on the level of gene expression desired. Inducible promoter systems, which can be turned on or off in response to certain drug or compounds, also are being investigated, as are self-regulating expression systems that provide very high levels of expression. The variety of gene expression systems available and under development expands the potential applications of gene therapy to new indications.

10.8. Routes of Administration

Even after the gene, vector and expression system have been optimized, the route through which a gene therapeutic is administered influences the therapeutic effect of the product.

10.9. Potential Benefits of Gene Therapy

A primary benefit of gene therapy is the ability to correct the underlying cause of genetic diseases. To date, the majority of therapeutics available for diseases such as cystic fibrosis, hemophilia and Gaucher's disease only alleviates disease symptoms. The delivery of functional copies of the genes involved in these diseases provides a mechanism through which the disease may be corrected at the molecular level. Gene therapy also holds the potential to provide patient-friendly treatment regimens for a variety of diseases. Today, patients with hemophilia, diabetes and other diseases that are treated by the administration of therapeutic proteins must take daily or weekly injections in order to manage their disease. This is because proteins exist in the blood stream for a limited period of time before they are degraded or eliminated. Because DNA is more stable and functions inside the cell, the delivery of therapeutic genes may result in long term expression of therapeutic proteins. Dosing regimens for gene therapy products are likely to require treatment every few weeks or every few months. In the area of cancer therapeutics, gene therapy has the potential to eliminate cancer cells without damaging normal, healthy tissue. In clinical studies to date, cancer gene therapies have been shown to be well tolerated and have a good safety profile. Cancer gene therapies may provide therapeutic alternatives to patients whose disease does not respond to standard treatment regimens. Additionally, cancer gene therapies may enable patients to maintain a better quality of life during treatment than do chemotherapeutic agents or radiation therapy. As the genomics revolution identifies new genes and as gene delivery technologies continue to evolve, the role of gene therapy in the treatment of a wide variety of diseases is likely to grow.

References

[1] Strachan, T. and Read, A.P. (1996) Gene therapy and other molecular genetic based therapeutic approaches in *Human Molecular Genetics,* Bios Scientific Publishers, Oxford.

[2] Wilkinson, G.W.G., Darley, R.L. and Lowenstein, P. (1994) Viral vectors for gene therapy, in *From Genetics to Gene therapy*, Bios Scientific Publishers, Oxford.

[3] Wolff, J.A. (1994) *Gene therapeutics*, Birkhauser Boston, Boston.

Chapter XI

Human Genome

11.1. Introduction

It was in December 1984 when during a workshop intended to evaluate the current state of mutation detection and characterization and to project future directions for technologies to tackle the prevailing technical limitations, the scientists discussed about the human genome analysis and first step in the direction was nucleotide sequencing of the entire human genome. The workshop was being sponsored by the US Department of Energy. In the workshop, growing roles of existing DNA technologies especially the emerging gene cloning and sequencing technologies were discussed. It was realized that existing technologies are in use for about a decade and mostly individual scientists are engaged in cloning and characterization of single genes which looked to be wasteful of manpower and also of research resources. It was also debated that such methodologies were incapable of determining mutations with good sensitivity. The scientists became of the view that an exhaustive, complex and expansive project for complete nucleotide sequencing of the human genome should be undertaken. Subsequently, a report on Technologies for detecting heritable mutations in human beings initiated the idea for a dedicated human genome project by the US Department of Energy (DAE). Afterwards, in 1986, DAE sponsored an international meeting in Mexico to assess the desirability and feasibility of ordering and sequencing DNA clones representing the entire human genome. All the scientists concluded that such a project was feasible and would be an outstanding achievement in biology.

Afterwards, DAE set three major objectives for the project which should be implemented: generation of refined physical maps of human chromosomes; development of support technologies and facilities for human genome research; and expansion of communication networks and of computational and database capabilities. Implementation of this program began with a small number of pilot projects. Simultaneously other US organizations initiated their own studies. In 1988, mainly with the support from the US Office of Technology Assessment and National Research Council and US National Institutes of Health (NIH), Office of the Human Genome Research was set up which was later renamed as the National Center for Human Genome Research. The aim of the center was to coordinate NIH genome activities in cooperation with other US organizations. In 1988, the US Congress also officially approved 15 years US human genome project commencing in 1991 and cost of the project was estimated to be nearly $3 billion.

In US, Department offices of Energy Research and of Health and Environment Research and Human Genome program, Department of Health and Human Services including NIH, Public Health Service and National Center for Human Genome Research were the nodal agencies. The DOE funded to number of laboratories like the Lawrence Berkeley National Laboratory, the Los Alamos National Laboratory, the Lawrence Livermore National Laboratory. Thereafter, in Europe countries like Germany, France, Italy, Denmark, The Netherlands, the United Kingdom also started the projects. Nordic countries proposed a cooperative genome initiative to contribute jointly to the human genome mapping efforts. The countries like Australia, Canada, Japan, Korea, New Zealand also started national genome projects. That's way, it became truly an international project. Later, Human Genome Organization was established to coordinate the different national efforts, facilitate exchange of research data, public debate etc. Three centers of Human Genome Organization (HUGO) were established; HUGO Europe (London); HUGO Americas (Bethesda); and HUGO Pacific (Tokyo).

It was considered that DNA technologies available would not deliver the desired results within the stipulated time period and budget. There was need for improved technology. Fortunately, improvement in the DNA sequencing technology became dramatic so that DNA sequencing costs have fallen effectively however without increase in the sequencing efficiency.

11.2. Human Genetic Maps

Classical genetic maps for experimental organisms for example, mouse, *Drosophila* were available since several years and have been refined successively. These maps were prepared by crossing different mutants to determine whether the two gene loci are linked or not. In human, preparation of genetic map was considered based on polymorphic markers which were not necessarily related to disease or to genes. The markers showed Mendelian segregation and were polymorphic, therefore, recombinants could be scored in a reasonable percentage of meioses and genetic map could be prepared. However, the main problem was of unavailability of polymorphic markers. The genetic markers were protein polymorphisms, particularly blood group and serum protein markers. Afterwards, in 1980s, construction of the human genetic maps was considered using restriction fragment length polymorphisms (RFLPs) solving the problem. The genetic maps helped in studying the nature of recombination in humans. Besides, it also helped for determination of gene localization. Any gene could be placed on the genetic map and its chromosomal localization could be obtained after polymorphic typing. Linkage analyses could be used in families segregating a disease gene, enabling the genes to be mapped for many inherited conditions. The genetic maps also helped in gene cloning. The linkage analysis are also helpful in prenatal or presymptomatic diagnosis of inherited disease genes by having many DNA markers in the vicinity of the disease locus.

There was first report of human genetic map in 1987 based on the use of 403 polymorphic loci including 393 RFLP markers. A major limitation was the reliance on RFLP markers which are not very informative markers and are difficult to type.

Afterwards, high resolution genetic maps have been constructed through the use of microsatellite markers. These markers were more polymorphic than RFLP markers. Hypervariable mini satellite polymorphism could be considered as an alternative but only in limited way. These micro satellites are distributed near the telomeres and not widely distributed throughout the genome. These micro satellites also called as short tandem repeat polymorphisms or SRTPs, are abundant. Thereafter, in 1992, second human genetic map was published selecting polymorphic CA/TG repeats, mapping them to specific chromosomes by typing panels of human- rodent somatic cell hybrids and performing statistical linkage analyses on markers from individual chromosomes. A total 813 markers were organized into 23 linkage groups.

Afterwards, maps have been produced with increasing numbers of genetic markers, especially the microsatellite markers and ever increasing resolution. Using different marker sets, many different genetic maps have been constructed. However, whether all the genetic maps have not used the same sets of reference families and relationship between different markers used in genetic maps is not clearly known. That's why, importance of international collaboration has been realized. Nowadays, the most widely distributed set of reference families is that deriving from the Centre d'Etudes du Polymorphisme Humaine (CEPH) collaboration of more than 100 independent laboratories. Nowadays, integrated genetic map is prepared.

11.3. Human Physical Maps

A physical map will consist of 24 component maps corresponding to 24 chromosomes. The first human physical map was obtained using cytogenetic banding technique which distinguished not only the different chromosomes but also discrimination of different sub-chromosomal regions. It became helpful as a framework for ordering the locations of human DNA sequences by chromosome *in situ* hybridization techniques. Other maps have also been obtained by mapping natural chromosome breakpoints using panels of somatic cell hybrids containing fragments derived from translocation and deletion chromosomes, or by mapping artificial breakpoints using radiation hybrids. Howcvcr, all these physical maps had limited resolution for large part of the genome. In fact, higher resolution maps are required and if single base pair resolution is desirable, that is the complete nucleotide sequence of the genome.

A complete contig (contiguous) map of the DNA of each of the 24 different types of human chromosome may correlate different DNA clones to define a series of partially overlapping DNA molecules covering the entire length of a chromosome. It would comprise all the DNA without any gaps. Identification of overlaps between the DNA segments of different clones can be done by a variety of different procedures. Yeast artificial chromosome (YAC), bacteriophage lambda, P1, PAC etc. have been used as vectors. Most of the effort in constructing contig maps of individual human chromosomes has involved isolation and characterization of YAC clones from total genomic YAC libraries. For construction of YAC clone contig map for a specific chromosome, Sequence Tagged Site content mapping has been very useful.

Scientists at the CEPH laboratory in Paris assembled contig maps of all 24 chromosomes in parrallel by random clone fingerprinting. Overlaps between YAC clones were identified using the following three methods:

1. Repetitive DNA fingerprinting: Clone fingerprints of all 33,000 YACs were produced by detection of restriction fragments containing medium repeat sequences (THE and LINE-1 elements).
2. STS content mapping: The STSs were selected as highly polymorphic microsatellite markers (STRPs). A total 2100 such markers were used to screen partially or totally 33,000 YACs.
3. Alu-PCR probe hybridization: The Alu-PCR reaction was used to derive clone specific DNA fragments for hybridization assays to test the relatedness of different clones.

Ultimately, Public Human Genome project scientists and Celera Genomics published the entire human genome sequence in February, 2001 issues of Nature and Science, respectively. Due to change in the policy of the US Government, now complete genome sequence is available on the internet (www.ornl.gov/hgmis/project/journals/journals.html).

The sequence has been found to be magnificant and is being considered as unprecedented resource and basis for research and discovery. It has been predicted that amount of the data is so huge that even if thousands of scientists work throughout the century, it will not be complete. Of course, the prediction is on the basis of available technology. It has been considered that research on human genome sequences will have diverse practical applications and impact upon how we feel ourselves and our place in the network of life around us.

After the entire sequence availability, scientists have estimated the total number of genes in human genome to be nearly 30,000 to 35,000. Some others have estimated up to 45,000. It is certainly much lower number than the earlier predicted number, nearly 100,000 or even more. Now if we compare the number of genes found in other organisms having much lesser complexity like *Caenorhabditis elegans* (roundworm) has 19099 genes, it is not much difference in number compared to complexity. It has been explained that genetic key to human complexity is not in number of genes but is, how gene parts are translated to synthesize different protein products. Now the earlier concept of one gene one protein got changed, and it is estimated that from one gene, even there may be synthesis of up to twenty proteins depending upon the age of the human. It is due to the process of alternate splicing. It has been found that in a sequence, a portion of the sequence may act as intron for one open reading frame and the same portion or a part of it may act as exon for another open reading frame. Many times, the exons may get ligated in different numbers and/or sequence forming different open reading frames. Besides, thousands of post translational chemical modifications are made to proteins, and regulatory mechanisms controlling these processes add to the complexity. According to estimates, there is synthesis of nearly 500,000 proteins from these 30-35,000 genes.

In constructing the sequence draft published in February, 2001, sixteen genome sequencing centers produced over 22.1 billion bases of raw sequence data, comprising overlapping fragments totaling 3.9 billion bases providing seven fold coverage of the human genome. Out of the total sequence data, it is considered that nearly 30 % data is of high quality, finished sequence with eight to ten fold coverage. The accuracy of this 30 % data is estimated to be nearly 99.99 % having few gaps. All the sequence data have been loaded on the web site for free availability to the public (www.ornl.gov/hgmis/project/journals/ sequencesites.html).

The entire working sequence draft has been completed in 2003. In 2003, there was also 50th anniversary of Watson and Crick's publication of double helix DNA structure. There was great celebration of 50th anniversary of the publication with the entire working sequence draft. However, still much remains to be deciphered. Final complete human genome sequence was presented on October 22, 2004. On the basis of the available data, the following points have been highlighted in the human genome sequence:

- Human genome contains 3164.7 million nucleotides.
- The average sized gene has about 3000 nucleotides. Although the size varies much. It is pertinent to mention that largest known human gene is dystrophin having nearly 2.4 million bases.
- Total number of genes in the human genome has been estimated nearly 25,000. It is pertinent to mention that this number is much lower than the previously estimated number, 80,000 to 100,000. Some scientists predicted the number up to 140,000. These estimates had been based on extrapolations from gene rich areas, whereas, the fact remained that human genome contained gene rich and gene poor areas.
- The precise number of human genes might never be tallied. Geneticists gathered in 2003 at a Cold Spring Harbor Laboratory meeting in New York said that they are no closer to a final tally. According to Phil Green, we do not know the true gene number for any organism and certainly not for humans. In 2003, Genetic database Ensembl predicted 24,847 genes in humans.
- The gene prediction programs which trawl through DNA for landmark sequences characteristic of a gene are unreliable since they often erroneously pick up pseudogenes, copies of real genes that have become defunct.
- On the other hand, the programs can miss genes carrying variations in their landmark sequences and are completely flummoxed by unconventional cases like tiny genes, overlapping genes or smaller genes hidden within larger genes. As per Gerald Rubin, a renowned geneticist, no gene predictor will ever get the number of genes right.
- As per Jean Weissenbach, Director of the sequencing centre Genoscope in Evry, France, there will never be a final number for the humban genes. It is possible that every person has a different number of genes, as mutations have eliminated minor genes from their genome.
- Almost all (nearly 99.9 %) nucleotide bases are exactly the same in all the persons.
- Still the functions are unknown for over 50 % of the discovered genes.
- Less than 2 % of the genome codes for proteins.

- Nearly 50 % portion of the genome is comprised of repeated sequences that do not code for any protein. These repeated sequences have also been called as **'Junk DNA'**.
- It has been considered that although these repeated sequences have no direct functions but they shed light on chromosome structure and dynamics. It has been considered that these repeats reshape the genome by rearranging it creating entirely new genes and modifying and reshuffling existing genes.
- During the last fifty million years, a dramatic decrease seems to have occurred in the rate of accumulation of repeats in the human genome.
- The human genome's gene dense 'Urban centers' are predominantly composed of the DNA building blocks G and C.
- In contrast, the gene poor 'Deserts' are enriched in the DNA building blocks A and T. GC and AT rich regions usually can be seen through a microscope as light and dark bands on chromosomes, respectively.
- Genes appear to be concentrated in random areas along the genome with vast expanses of non-coding DNA in between.
- Stretches of up to 30,000 C and G bases repeating over and over often occur adjacent to gene rich areas forming a barrier between the genes and the junk DNA. These CG islands have been predicted to have role in regulating the gene activity.
- Chromosome 1 has maximum number of genes and that is 2968. Whereas, chromosome Y has only few genes and the number is 231.
- The human genome is unique in having random distribution of gene rich areas. It has been found that many other organisms genomes are more uniform with genes evenly spaced throughout.
- Humans have on an average three times as many kinds of proteins as the fly or worm. This is because of mRNA transcript's alternate splicing and chemical modifications to the proteins. Due to alternate splicing and/or chemical modifications, different protein products may be synthesized from the same gene.
- Humans share most of the same protein families with worms, flies and plants. However, the number of gene family members is expanded in humans especially in proteins involved in development and immunity.
- Human genome has much greater portion of repeat sequences which is nearly 50 %. Other eukaryotes have lower percentage of the repeat sequences. For example, *Arabidopsis* (mustard weed) has 11 % repeat sequences. The worm and the fly have 7 % and 3 % repeat sequences, respectively.
- Humans appear to have stopped accumulating repeated DNA sequences over 50 million years ago. However, no such decline has been found in rodents. This difference may be due to some of the fundamental differences between hominids and rodents. It is pertinent to mention that gene estimates are similar in these species. The scientists have proposed many theories to explain evolutionary contrasts between humans and other organisms which also include life span, litter sizes, inbreeding, genetic drift etc.

11.4. Variations and Mutations

1. About 1.4 million locations have been identified having single nucleotide differences in humans. This information seems useful to revolutionalize the processes of finding chromosomal locations for disease associated sequences and tracing human history. The human sequence project made the biologists acquainted with the importance of single nucleotide polymorphism (SNP). The SNPs define differences among people, predispose a person to disease, and influence a patient's response to a drug. It is now at least possible to build microarrays that can target individual SNP variations, as well as making deeper comparisons across the genome.
2. The ratio of germline (sperm or egg cell) mutations is 2: 1 in males versus females. Several explanations have been given for justifying higher mutation rate in the male germline that also include the greater number of cell divisions required for sperm formation than for eggs.

11.5. Gene Predictions

Gene predictions are made by computational algorithms based on the recognition of gene sequence features and similarities to known genes. It has been estimated that *Saccharomyces cerevisiae* has 6034 genes and the size of the genome is 12.1 million bases. The number of genes per million sequenced DNA bases (gene density) has been found to be 483. In *E. coli*, number of genes has been predicted to be nearly 4200 whereas the size of the genome 4.6 million bases. The gene density in *E. coli* has been predicted to be 932 which is nearly double than *Saccharomyces*. In *Caenorhabditis elegans*, the number of genes has been predicted to be 19099 and genome size, 97 million bases. The gene density has been predicted to be 197, much lower than yeast and *E. coli*. In *Arabidopsis thaliana* and *Drosophila melanogaster*, the number of genes has been predicted to be 25,000 and 13061, respectively. The sizes of the genome have been reported to be 100 and 180, respectively, whereas the gene density as 221 and 117, respectively. In humans, gene density has been reported to be 12 only. However, it is pertinent to mention that these data need further confirmation including characterization of their protein products and functions.

11.6. Applications, Future Challenges

Deriving meaningful knowledge from the DNA sequence will define research to inform understanding of biological systems. This job will require expertise and creativity of tens of thousands of scientists from varied disciplines in both the public and private sectors worldwide.

The draft sequence has impact on finding genes associated with diseases. At present, over 30 genes have been pinpointed and associated with breast cancer, muscle disease, deafness and blindness. Job of finding the sequences responsible for common diseases like cardiovascular disease, diabetes, arthritis and cancers is being carried out by using human variation maps. These studies provide focused targets for the development of effective new therapies.

Having this sequence will enable the workers a new approach to biological research. In the past, researchers studied one or few genes at a time. With whole genome sequences and new high throughput technologies, they can approach questions systematically and on a grand scale. The scientists can study all the genes in a genome, for example, or all the transcripts in a particular tissue or organ or tumor, or how tens of thousands of genes and protein work together in interconnected networks to orchestrate the chemistry of life.

Rapid and accurate technologies are needed for differential sequencing to detect rare one nucleotide substitution in thousand nucleotides to distinguish different genomes in order to characterize polymorphisms. At present, mass spectrophotometry techniques are available to perform 250,000 assays a day for detection of single mutation at an error rate of one in 10,000. At present, the scientists are engaged to assemble the database required for associating complex disorders with sets of common alleles. Many concepts in biology are about to change. Reference genomes along with various machine learning algorithms developed are also bringing microarray technologies to full power. It is expected that in near future, annotation of the human genome will be complete and probes for every gene will be arrayed on a solid phase substrate of a few square centimeters. The microarray technology does better compared to other current methods to monitor and characterize change, whether due to normal cell growth and development, progression toward disease, or response to exogenous ligands. It is expected that diversity would be characterized with precision and breadth that may revolutionize medicine during the coming years. It is expected that in next few years, there will be emergence of high throughput technologies for profiling proteins in various states of modification. There is need of bioinformatics tools for information processing.

Bioinformatics is a branch of computing concerned with the acquisition, storage and analysis of biological data. Bioinformatics offers much efficiency through mathematical modelling. In the search for new drugs, it promises new ways to look through data mining. In fact, it is the only practical way of making sense of the huge data. The progress in bioinformatics occurred after development of powerful software for computer aided design and computational fluid dynamics. Nowadays, Biotechnology industries are investing for computer modelling, data mining and high through put screening to discover drugs more efficiently.

Bioinformatics has actually become a huge platform of technologies covering computer architecture like workstations, servers, supercomputers, storage and data management systems, knowledge management and collaboration tools and the life science equipment needed to tackle biological samples. In fact, the credit of computerization of genetics on such a huge scale goes to the invention of DNA microarrays, sequencing of human genome and dramatic increase in computing power. Microarrays, more commonly known as gene chips, are to the genetic revolution of today what microprocessors were to the computer revolution. Microarrays have changed the difficult task of screening genetic information into an automatic routine that expoits the tendency for the molecule that carries the template for making the protein, messenger RNA, to bind to the DNA that produces it. Gene chips contain thousands of probes each imbued with a different nucleic acid to bind with messenger RNA. The resultant bonds fluoresce under different colors of laser light showing which genes are present. Microarrays measure the incidence of genes and their abundance. The importance of gene chips is clear from the fact that with in few years, gene chips have gone from experimental novelties to tools of the trade.

References

[1] *Human Genome News* (2001) Vol.11 (No.3-4, July 2001). U.S. Department of Energy Office of Biology and Environmental Research.
[2] Donis-Keller, H., Green, P., Helms, C. (1987) *Cell*, 51, 319-337.
[3] McKusick, V. (1989) *Genomics*, 5, 385-387.
[4] Weissenbach, J., Gyapay, G., Dib, C., Vignal, A., Morissette, J., Millasseau, P., Vaysseix, G. and Lathrop, M. (1992) *Nature*, 359, 794-801.
[5] Murray, J.C., Buetow, K.H., Weber, K.L. (1994) *Science*, 265, 2049-2054.
[6] Dausset, J., Cann, H., Cohen, D., Lathrop, M., Lalovel, J.M., White, R. (1990) *Genomics*, 6, 575-577.
[7] Cohen, D., Chumakov, I., Weissenbach, J. (1993) *Nature*, 366, 698-701.
[8] Strachan, T., Read, A.P. (1996) Human Molecular Genetics, Bios Scientific Publishers, Oxford.
[9] Pearson, H. (2003) *Nature*, 423, 576.
[10] *www.economist.com*

Chapter XII

Proteomics and Its Promise

12.1. Introduction

In February, 2001, there was a monumental achievement when Celera Genomics and Human Genome Project each published their sequences of the human genome.

After that, people started asking what comes next after the sequence of the human genome? Researchers said –Proteomics that is the study of proteins coded by genes. Proteomics may be considered as the systematic cataloging, separation, and study of all the proteins produced by genes with in each cell as well as the complex interactions among proteins that ultimately result in health or disease. The another term coined is Proteome which according to Joshua Lederberg, a Nobel laureate refers to the whole body of diverse proteins found in an organism. According to Joshua Lederberg, the term 'Proteomics' and 'Proteome' only came into use between 1995 and 1998 by analogy with genomics and genome, and are still not in standard dictionaries. The term, 'Proteome' defines the entire protein complement in a given cell, tissue or organism. The proteome research also assesses protein activities, modifications and localization, and interaction of proteins in complexes. It is very much a technology driven enterprise. The proteomics research will boost our understanding of systems level cellular behavior through study of global patterns of protein content and activity and changes during development or in response to disease. It is also expected that clinical research will be benefitted from proteomics by both the identification of new drug targets and the development of new diagnostic markers.

According to Professor Brian T. Chait of Rockfeller University, proteomics addresses the question of what proteins do in a cell in a global, integrated way. It is the study of where each protein is located in a cell, when the protein is present and for how long, and with which other proteins it is interacting. It means looking at many events at the same time and connecting them.

According to J. Lederberg, we know tens of thousands of proteins at this stage but just as the periodic table enabled us to say that there was a larger number of elements that had yet to be discovered, that is true currently about proteins.

If we talk of the protein synthesis, we all know that genetic information for the sequence of amino acids in the proteins is located in DNA. From DNA, there is formation of messenger RNA transcript which is directly involved in protein biosynthesis.

Proteins are composed of amino acids which are arranged according to particular sequences which correspond to the sequence of nucleotides in the gene. After synthesis, many proteins are also chemically modified. At this stage, the protein essentially has all the information necessary to adopt its three dimensional structure.

According to Professor Hochstrasser of University of Geneva, proteins are the workhorse of the cell and all proteins work together in a complex network to give function. According to John Richards from CALTECH, genes are the blueprints for information required for life, but proteins expressed in different cells are the dynamic machines responsible for function.

As has been mentioned in the chapter on human genome, from the human genome sequences, it has been estimated that there are 30,000 to 35,000 genes. How do we manage to be so complex with so few genes? It is apparent that our complexity is tied not to the number of genes but rather to the complexity of their products, the proteins. If we compare the number of genes in fruitfly, roundworm and human, it is nearly 13000, 19000 and 30,000, respectively. All the three organisms share many homologous genes. It has been predicted that the key to each organism's uniqueness lies in the fact that each gene may produce more than one protein anywhere from six to twenty or even more. Now the older concept of one to one ratio of genes to proteins no more exists. The human genome diversity is tremendous and is several order of magnitude greater than the genome diversity. Each gene normally expresses five-six proteins and up to twenty with increasing age. According to estimate, a human may express up to 500,000 proteins. However, till now only about 80,000 proteins have been identified and out of these 80,000 proteins only small number of proteins have been studied in detail. In fact, the pace of discovery in proteomics has accelerated and this field is entering in an exciting new era.

How do genes produce such a diversity of proteins? This is due to the way instructions for protein synthesis are transmitted from DNA to the cytoplasm. Information for making proteins in DNA is arranged in blocks called exons which are interrupted by other sequences called as introns whose function is not understood yet. The cell edits the RNA copy assembled on the DNA by removing introns, a process called as RNA splicing. In fact, splicing is a more complex process than a simple linear editing process.Cells can use alternative splicing schemes to generate a variety of messages for a given sequence of DNA. For example, a linear sequence of exons 1,2,3,4,5,6 can not only generate mRNAs whose sequence is 1-2-3-4-5-6 but also other sequences such as 1-2-4-6, 13-5-6, 1-4-5-6 etc. It is called alternative splicing. Sometimes, a portion of the gene which acts as intron for one protein may act as exon in toto or partially for another protein. Thus using alternative splicing, the same gene sequence can result in a number of different proteins. Protein diversity also arises because proteins are frequently modified after synthesis. For example, two proteins may be covalently linked through –S-S- linkage to form a dimer. There may be phosphate or sulfate addition on the protein. In some cases, cleavage occurs for getting active proteins and example may be considered of zymogens. Thus, by alternative splicing schemes and post translational changes, few genes may give large number of proteins.

Human proteomics is considered to be important for knowing more about diseases. The disease is a malfunction of physiological pathways. The fundamental underlying cause of disease is that proteins which do all the jobs in the cell, get out of kilter. Proteins have also been considered to be a type of nano-machines. It is being hoped that proteomics may help in drug targeting. Since, drugs work by correcting protein malfunction by increasing or decreasing their amounts or by altering their interactions. Therefore, by studying proteins, one may expect not only to understand the nature of the disease but also learn to design drugs that are more effective than drugs developed. In new drug development, there are number of crucial points in which proteomics can guide scientists. According to Michael Silber of Pfizer, the first crucial point would be the identification and selection of good targets compared to superfluous ones. A drug target is a protein that might be inhibited or activated to produce a therapeutic effect. The drugs are as good as the targets selected. This will certainly help in the discovery of new and more specific medicines. The next point is how many proteins or more precisely specific sites in proteins can be targets for therapeutically valuable compounds. At present, the estimate is variable ranging from 1000 to 20,000. However, currently pharmaceutical companies are working with 400-500 targets only, and many of which are receptors of only one particular type. Therefore, the field is potentially wide open. However, it is essential to determine which targets are most urgent, most important and most accessible to research. However, it is not too simple to work. One most important complication is variants of the same protein. For example, cytochrome P-450 is responsible for metabolism of many drugs in the body. It may explain why the same drug may affect some patients differently. Variations in the same protein are known to be involved in individual susceptibility to some diseases. Therefore, the number of targets is actually amplified by the number of variants of each gene and the proteins produced by it. In fact, the number of targets is probably even larger than the total number of proteins produced because proteins interact with each other and form larger protein complexes that can be targeted specifically.

One way of study is to compare homologous genes and proteins among species to understand function. If one finds a protein in mice that has a specific function, it may do the similar type of function in human and other animals. An example is of Pax 6 gene that affects the eye development. If this gene is compromised in mouse, in human, they all are blind. Such type of data may provide strong clue about a certain gene and its proteins might be worthwhile to pursue as target. In fact, the validation of targets remains exceedingly complex at this stage. In fact, it is necessary to know much more than the sequence information or even the presence of unique protein signatures. It is really necessary to understand protein's structure and function, and to examine the interplay of the multiple targets being up and down regulated. In fact, proteomics is the primary tool for really checking what the drugs do, whether they manipulate a target? Proteins are too numerous, diverse, and interactive to be studied by a single technique. To study proteomics, the approaches and tools developed to discover and mine genomic information are being adapted, and sometimes combined with the older, established methods. The scientists are engaged in discovering the limits of the available technology and are working to find ways around those by developing new approaches and technologies. One of the challenges is to identify how proteins are related in order to understand the role of particular proteins. It is necessary to know which proteins interact with which others and what pathways they follow. Proteins are dynamic moieties. Therefore, it is essential that there should be much advancement in technology to meet the challenges of proteomics.

After evaluation of the future prospects, many academic scientists, and pharmaceutical and biotechnology companies have come together to work as a team, and many other companies are doing so at a rapid rate. For example, Anderson's company, Large Scale Proteomics, Biosite Diagnostics are collaborating to develop protein chip arrays, also called antibody arrays as tools to measure large number of proteins in biological samples. It is expected that such chips will be the preferred technology for high volume applications of clinical research, diagnostics and toxicology. Celera Genomics is comparing proteins expressed by comparing both types and amounts in healthy and diseased tissue as well as drug treated and non-drug treated samples in order to find targets for a range of diseases. Myriad Genetics, Hitachi, and Oracle have joined hands with an investment of half a billion dollars to identify all human proteins and all their interactions. Incyte Genomics and Genicon Sciences are also working to detect infinitesimal levels of proteins present in tissues using Incyte's antibody array products. Several companies are focusing on ways to organize, manage and possibly mine the available information. LSP's Human Protein Index (HPI) Database aims inventory proteins occurring in all major human tissues. On the basis of results of protein characterization from Large Scale Proteomics's high throughput Mass Spectrophotometry facility, HPI currently covers protein products of nearly 18,000 human genes in 137 different human tissues. The scientists are in the process of mining the data to determine what proteins are characteristic of different regions of the brain, heart muscle, liver and kidney etc. It is being considered to be the beginning of the translation of results of this field into medical benefit.

A typical cell may contain thousands of proteins at any time. Proteins play a variety of roles in the cell. A large class of proteins called enzymes plays an essential role in catalyzing all the biochemical (anabolic as well as catabolic) reactions in the cell. Other proteins such as actin play a structural role in the cell and give shape to the cell. Some proteins also help in forming organelles in which different cellular functions are partitioned. Some proteins bind with nucleic acids and other cellular constituents. Some proteins like insulin act as hormones. Some proteins also work as antibodies. Some proteins are involved in transportation of other molecules like hemoglobin transports oxygen.

12.2. Four Steps of Protein Formation

The following four steps have been considered to be important in protein formation:

1. **Transcription and Translation** - All the genetic information in the cell is carried in DNA which resides in the nucleus. To make proteins, information from DNA is abstracted by making a copy of the appropriate part(s) of the genome by a process called as transcription. Transcription and some steps (post translational changes) immediately following it result in the formation of mRNA which has genetic information for the sequence of the amino acids in the protein to be synthesized and

therefore serves as template. For the purpose, mRNA leaves the nucleus for the cytoplasm where it binds with the ribosome, an organelle present in the cytoplasm and responsible for protein biosynthesis. On the ribosome, protein synthesis occurs in the presence of mRNA, a process called as translation. Once, mRNA binds with the ribosome, a smaller size RNA called as transfer RNA (tRNA) starts to bring amino acids one at a time to the ribosome like making a necklace with one bead at a time. The sequence present in the mRNA dictates which of the tRNA binds at any specific point. The sequence of the nucleotides in DNA via mRNA intermediary, directs the synthesis of protein with a precise and predetermined sequence. Specific signals occur on the mRNA for the start and termination of the protein synthesis.

2. **Protein Folding: Generation of Functional and Active Proteins -** Proteins have specific three-dimensional structure. Many proteins are globular whereas others are fibrous in shape. Few proteins may also have other shapes too. It has been postulated that physiological environment plays a role in protein folding. In some proteins, there is exposure of the hydrophobic region on the surface which helps in binding with the hydrophobic regions like with the membrane. There is involvement of covalent bonds and a variety of non-covalent bonds interactions between the amino acids for folding to give a specific three dimensional structure.
3. **Post Translational Modifications: Protein Diversification by the Addition of Sugars, Phosphates and Other Molecules -** There are post translational modifications in many proteins made by the addition of sugars (glycosylation), phosphate (phosphorylation), sulfate and few other molecules. These post translational modifications are important since these are essential for modulating the function of the protein. Some proteins get rendered active or inactive by these modifications. Few proteins acquire immunogenic properties due to glycosylation.
4. **Protein Protein Interaction: Protein Function through Binding or Formation of Complexes -** Although most of the proteins act by themselves. Some proteins are active only when they are in a complex either by binding with other molecules of self (forming dimer, tetramer or so) or with other proteins or cellular constituents. For example, when four molecules of hemoglobin are bound together to form tetramer only then it is active for the transport of oxygen. Proteins form complexes by binding along surface clefts created by folding in a particular fashion as well as by ionic and other interactions. Multienzyme complexes may also be considered as other examples of protein protein interaction. Study of protein protein interaction is considered to be an important area of research.

12.3. Proteomics for Diagnosis and Prognosis of Disease

Proteins are being used as bio-markers in medicines for diagnosing and staging disease and for determining prognosis. If more sensitive, efficient techniques are developed to discover and analyze proteins, there will be improvement in understanding molecular basis of disease, and subsequently better methods of diagnosis, treatment and prevention of diseases may be evolved out. Prognosis is more important than diagnosis. If one is able to identify crucial proteins, it may help to determine a patient's prognosis and also select a treatment.

For example, it has been reported that troponin-I, a cardiac specific blood protein, may predict the risk of death in the next 42 days of a heart attack patient. If the level of troponin-I is below 0.4 nanogram per milliliter when the person comes in emergency with chest pain, risk of dying with in 42 days is less than 1 %. If the level is greater than 9, the risk is nearly 8 %. Therefore, just by measuring one protein in the blood but of course the right one, the physician may decide whether the patient should go home or stay in the Intensive Care Unit to do something about the coronary artery disease.

Identifying key proteins in disease can also guide drug development and treatment selection. Hamm et al. (1999) studied troponin-T, a close relative of troponin-I which has also been investigated in heart attack patients having obstructed coronary vessel. They treated patients with a therapeutic antibody to prevent platelet aggregation and obstruction of blood vessels. They found that antibody treatment had no particular effect in patients with low levels of troponin-T. However, in patients with high levels of troponin-T who were at high risk for recurring myocardial infarction, the antibody was quite effective in preventing subsequent myocardial events. This study implies that it is appropriate to limit treatment with the antibody only to those patients who have high levels of troponin-T because they are the ones who would benefit from the drug. This ability to match individual patients and specific drug treatment has a tremendous future potential.

There are other examples where specific proteins or enzymes have proven very useful for treatment or prognostic purposes. However, the major obstacle to wider applications of proteomics in clinical medicine is the slow speed of protein separation and identification in the laboratory and difficulties in dealing with more than very few proteins at a time. We need faster and better ways to analyze proteins on a large scale.

Efforts are being done for automating the methods of protein separation and analysis with large databases which offer the opportunity to compare the generated data with a large body of existing information for rapid comparison, characterization and possible identification. Work in this direction is in progress in number of institutions.

As these and other approaches to make proteomics studies faster, better and cheaper at a very rapid pace around the world, scientists interested in clinical applications of proteomics look to a future where patient care will be directly linked to specific and individualized information. The ultimate goal is to be able to send a biopsy specimen of a suspicious lesion from an individual patient to the lab or even better, perhaps a saliva, or skin sample to determine whether it shows cancer, and if so what type, what stage and to which drugs it will be sensitive. The combination of information that will be obtained from genes and proteins is expected someday to make patient care more scientific and therapies more specific and effective as compared to today.

12.4. Analyzing Protein Structure and Function

Each human cell may contain several thousands proteins. It must be known where and how much of a protein is present and for how long, and with what protein it is interacting. Although, the structure of individual proteins and the shapes and topology of their interacting complexes are under active investigation in many laboratories, still new techniques are required to be developed to accelerate the pace of these investigations.

An important goal of proteomics is also to devise tools that will help scientists in analyzing cellular functions which they expect will lead to a better picture of normal processes as well as of disease mechanisms.

In a system as complicated as a multicellular organism, one must look at the entire system in an integrated way. The proteins work in complex partnerships with each other and various constituents of the cell. Therefore, it is necessary also to track the interactions and changes the proteins produce rather than viewing individual proteins in isolation.

Some scientists have studied protein interactions by tagging them with a sort of 'molecular velcro' which allows them to pull out the tagged protein together with its strongly associated partners. Afterwards, the proteins can be identified by mass spectrometry. One should also determine the type of the interacting complexes, the shapes of the individual proteins and modifications that regulate the function of the protein. This is a big job that requires many types of instruments. Much efforts are being done for determination of the three dimensional structure of proteins using X-ray crystallography and nuclear magnetic resonance spectroscopy.

Argonne National Laboratory, Lawrence Berkeley National Laboratory, Los Almos National Laboratory, Rutgers University, Scripps Research Institute, University of Georgia are doing work on proteomics with the support of National Institute of General Medical Sciences (NIGMS). These institutes wish to develop an inventory of all the protein structure families that exist in nature. These institutes are trying to streamline and automate X-ray crystallography and NMR spectroscopy. The Scientists are involved in organizing all known proteins into structural, or fold, families based on their genetic sequences. Afterwards, they would like to determine the structure of one or more proteins from each family. These scientists will like to use gene sequences to approximate structures of all other proteins. If the approach works, this program would certainly add significantly to our understanding of protein structure. Besides, by merging nine institutions, New York Structural Biology Center (NYSBC) has been established which is concentrating on using ultra high field nuclear magnetic resonance spectroscopy to determine protein structure and function. The research in NYSBC is mainly focussed on membrane proteins, high speed structure determination, flexibility and mobility of multi domain systems and screening for early lead drug candidates, all of these are key technologies to meet many of the challenges of proteomics. New York Office of Science, Technology and Academic Research is also funding the research in NYSBC.

Agouron and Vertex have claimed to develop anti-HIV drugs and Roche claimed to develop anti-influenza drug.

Like genomics, the grand scale of proteomics research is making it a community effort with the Human Proteome Organization (HUPO) playing an important role in coordinating proteomics projects worldwide. The size of information poses difficulty in data management and therefore publicly accessible databases are essential that use accepted standards to describe protein data, allowing data comparison and integration. Considering the money involved and grand scale of proteomics technologies, access and solutions must be found that allow the widespread use of protemics tools.

On the other hand, there are legal problems of Intellectual Property Rights (IPR). According to S. Leslie Misrock, a leading Attorney in USA having expertise in the field of IPR, if efficient means to resolve intellectual property disputes are not considered, the promise of proteomics can turn into a cataclysmic failure.

It is hoped that with the advancement of proteomics, more individualized medicines will be developed. It is not likely to be a single drug for one disease, but possibly a wide range of treatments for diseases based on their molecular fingerprints. It is hoped that proteomics is one area of great promise that scientists will mine for years to come to develop such medicines.

Like genomics, Human Proteome Organization (HPO) has been established which is playing a significant role in coordinating proteomics projects globally. The bulk of information produced poses challenges for Data management and necessitates publicly accessible databases that use agreed standards to describe protein data, allowing data comparison and integration. Besides, the money and scale of proteomics technologies restricts their access. Therefore, solutions must be searched that allow the widespread use of proteomics tools.

References

[1] *The Promise of Proteomics: Leading the way to 21st Century Medicine* (2001) published by the New York Academy of Sciences, New York.

[2] Antman, E.M., Tanasijevic, M., Thompson, B., Schactman, M., McCabe, C.M., Cannon, C.P., Fischer, G.A., Fung, A.Y., Thompson, C., Wybenga, D. & Braunwald, E. (1996) Cardiac specific troponin I levels to predict the risk of mortality in patients with acute coronary syndromes. *The New England Journal of Medicine* 335, 1342-1349.

[3] Hamm, C.W., Heeschen, C., Goldmann, B., Vahanian, A., Adgey, J., Macaya, C., Rutsch, W., Berger, J., Kootstra, J. & Simoons, M. (1999) Benefit of abciximab in patients with refractory unstable angina in relation to serum troponin-T levels. *The New England Journal of Medicine* 340, 1623-1629.

[4] Razvi, E. & Latterich, M. (2000) Technology Trends for Proteomics and Beyond. Technology for Proteomics and Beyond. *Drug & Market Development* 338-343.

[5] James, R. (2000) Differentiating Genomics Companies. *Nature Biotechnology* 18, 153-155.

[6] *Nature* (2003) 422 (No. 6928) The complete issue highlights proteomics.

Glossary

AIDS: A disease called Acquired Immunodeficiency Syndrome. It is caused by the virus, HIV (human immunodeficiency virus), a RNA virus, capable of destroying the immune system of the patient.

Alkaline phosphatase: It is an enzyme that catalyzes the hydrolysis of phosphate ester bond releasing inorganic phosphate at alkaline pH. In molecular biology, it has been used for removing phosphoryl group from the 5'-ends of the DNA strands. Molecular biology grade alkaline phosphatase from calf intestine and bacteria are available commercially. Bacterial alkaline phosphatase has optimum temperature 65°C whereas calf intestine alkaline phosphatase has optimum temperature 37°C.

Ampicillin resistant gene (Amp^r): It is a gene whose product causes ampicillin resistance to the host cell. The gene codes for a protein secreted into the periplasmic space of the bacteria where it catalyzes the hydrolysis of the β lactam ring of ampicillin thus destroying the antibiotic.

Antisense gene: The antisense gene allows the stable and permanent expression of antisense RNA which binds to mRNA and prevents translation. Increasing use of antisense gene construct containing ribozyme is being considered which may degrade the mRNA to which it binds.

Antisense RNA technology: From double stranded DNA, there is transcription of single stranded mRNA which is involved in protein biosynthesis for providing genetic information. If an antisense RNA (antisense to mRNA) is injected into the cell, it will bind with the mRNA and forms double stranded mRNA which will be no more active for protein biosynthesis. This technology is called as antisense RNA technology. It is useful for preventing the multiplication of pathogenic microbes present in the host cell. However, instability of the injected antisense mRNA in the host cell is the drawback with the technology.

Arbitarily primed –PCR (AP-PCR): In this technique of modified PCR, only one primer of relatively much smaller length as compared to the primers used in the conventional PCR is used which gets annealed to both the strands arbitarily and fragments of various lengths are randomly generated. The technique has been used for DNA profiling in biotechnology and in forensic medicine.

BAL-31: It is an enzyme that chews blunt ended double stranded DNA from ends. In molecular biology, it is used for ordered sequencing using dideoxy method of Sanger.

Base pair (bp): The binding of adenine nucleotide with thymine nucleotide (or uracyl nucleotide) and guanine nucleotide with cytosine nucleotide is called base pairing. Sometimes, due to the Wobble phenomena, other pairs can be formed.

Biotinylated probe: It is a non-radio active probe tagged with biotin. It works on the principle of biotin streptavidin specific binding. Streptavidin is used conjugated with alkaline phosphatase which may be detected using chromogenic substrate.

BluGene: It is a registered trade mark for biotinylated probe.

5-Bromo-4-chloro-3-indolyl-β-D-galactoside (X-gal, or bluogal): It is the chromogenic substrate of the enzyme β-galactosidase. It is colorless itself but after reaction with β-galactosidase, the product is bluish colored. Therefore, it is used for histochemical screening of the vetors. Bluogal is the registered trade mark of BRL.

cDNA: Single stranded DNA complementary to RNA synthesized from it using reverse transcriptase *in vitro*.

Chloramphenicol resistant gene (Cm^r or Cat): It is a gene whose product causes chloramphenicol resistance to the host cell. The gene codes for a tetrameric protein having a subunit molecular weight 23,000. The protein acts as chloramphenicol acetyl transferase catalyzing the transfer of acetyl group from acetyl CoA onto chloramphenicol forming hydroxy acetoxy derivative of chloram. The hydroxy acetoxy derivative of chloram is unable to bind to the ribosome unlike chloramphenicol. The expression of Cat is catabolite sensitive. There is 5 to 10 fold increased production if the bacteria is grown in a medium other than glucose.

Cloning site: If the recognition sequence of any restriction endonuclease is single in the entire sequence of the vector, it may be considered as the cloning site. However, generally, it is preferred that on cloning a DNA fragment in the cloning site, the recombinant should be easily detectable.

α-Complementation: Active β-galactosidase present in *E.coli* cells is a tetrameric protein which splits β-1,4 galactosyl linkage. *In vitro*, for screening of the recombinant vectors, colorless substrate X-gal also called bluogal (5-bromo, 4-chloro, 3-indolyl-β-D-galactoside) is used whose product is a blue chromophor thus allowing visual detection of β-galactosidase activity. Mutation in the 5'-region of the *E.coli lac* Z gene produces subunits incapable for assembly. Gene has been deleted from the host genome and mutated gene inserted into F episome. The protein synthesized from the mutant gene lacks amino acids 11-41 of the wild type protein. These subunits do not show β-galactosidase activity. The sequences incorporated into certain vectors are able to code for a short polypeptide corresponding to N-terminal region of the enzyme. The polypeptide after association with the subunits is capable of making active tetrameric enzyme restoring enzyme activity. This association is called as α-complementation.

Cosmid vectors: These are the modified plasmids that carry at least one *cos* site (DNA sequences required for packaging DNA into bacteriophage lambda particles).

Cos site: These are the DNA sequences required for packaging DNA into bacteriophage lambda particles.

3,4-Cyclohexenoesculetin-β-D-galactopyranoside (S-gal): It is a substrate of β-galactosidase enzyme used in the histochemical screening of the vectors (in place of X-gal). It is developed by Sigma –Aldrich and S-gal is their registered trade mark. It is autoclavable/ microwavable. On using S-gal as substrate with β-galactosidase, the hydrolyzed aglycone (non-sugar portion) reacts with the Fe^{+3} of added ferric ammonium citrate to produce intense black stain. As per Sigma, S-gal is more soluble in water than commonly used X-gal and is also not light sensitive unlike X-gal.

Cystic fibrosis: A genetic disease which eventually develop chronic pulmonary disease. The nature of the primary defect is a non-functional transmembrane conductance regulator (a nearly 170,000 daltons molecular weight protein) whose gene is located on the chromosome 7 in humans.

***dam* Methylase:** It is an enzyme that catalyzes the transfer of methyl group from S-adenosyl methionine on to N6 of adenine residue when present in a particular sequence, GATC. This sequence is present with in the recognition sequences of many restriction endonucleases and in most of the cases, after methylation, the site is not cut by the restriction endonuclease.

***dcm* Methylase:** It is an enzyme that catalyzes the transfer of methyl group from S-adenosyl methionine on to the cytosine residue forming 5-methyl cytosine when cytosine is present in a particular sequence, CCAGG or CCTGG. This sequence is present with in the recognition sequences of many restriction endonucleases and in most of the cases, after methylation, the site is not cut by the restriction endonuclease.

Denaturation (of DNA): This is the process of converting double stranded DNA into single stranded DNA using heat.

Deoxyribonucleic acid: It is a polynucleotide having deoxyribonucleotides as its constituents.

Diphenylamine: It is a chemical used for detecting and estimating deoxyribosyl moiety. In nucleic acids, it is used for estimation of DNA since deoxyribose is a constitutent of DNA.

DNA ligase: It is an enzyme that catalyzes the end to end joining of the DNA molecules. In molecular biology, it is used for cloning the DNA fragments in the vectors. It is also used for ligating specific linkers at the double stranded blunt ended DNA molecules.

DNA methylases: These are specific enzymes which catalyze the methylation of specific nucleotidyl residues using S-adenosyl methionine as the methyl transferring agent.

DNA polymerase I (of *E.coli*): An enzyme used to synthesize daughter DNA strand using DNA as a template *in vitro*. The enzyme is used during cDNA library formation. *In vivo*, it is considered to involve in DNA repairs. It has 3',5' exonuclease as well as 5',3' exonuclease activities in addition to DNA polymerase activity.

Dot blot hybridization: It is a technique used for detecting DNA molecule having a specific nucleotide sequence. Here, unlike southern blotting, DNA sample is directly applied on the nitrocellulose membrane and no electrophoresis is carried out. In genetic engineering, it has been used for screening mutants during oligonucleotide site directed mutagenesis.

Electroporation: It is a device of DNA transfer in the cell. In electroporation, sample of protoplast is pulsed with high voltage pulses in the chamber of electroporator. The chamber is cylindrical in form with distance of one centimeter between parallel steel electrodes and has a pulsed volume of nearly 0.3 ml. The pulse is applied by discharge of a capacitor across the cell. The decay constant of the pulse (time taken to decay to 1/e of the initial voltage) is on the order of 10 microseconds with a capacitor of 10 nF and a chamber resistance of one kiloohm. The resistance across the chamber is measured with an alternating current multimeter operating at one KHz. The field strength required is inversely proportional to the diameter of the protoplasts being treated and may vary a little from species to species in the strength required for a given size of protoplast. After transferring the protoplast, resistance is measured and is adjusted to about 1 to 1.1 kiloohm by adding $MgCl_2$ and afterwards is transferred in capped tube and heat shock is given at 45°C for 5 minutes followed by cooling to room temperature. DNA to be transferred is added to the protoplast followed by polyethylene glycol solution. After about 10 minutes, samples are transferred back in the chamber of the electroporator and pulsed three times at 10 seconds intervals with pulses of an initial field strength of about 1.4 KV/cm. Afterwards, pulsed protoplast is transferred in a suitable medium.

Exonuclease III: It is an enzyme that chews double stranded DNA molecule from 3'ends (if it is a blunt end). In molecular biology, it is used for ordered sequencing using dideoxy method of Sanger.

Gene augmentation therapy: It is the traditional approach to gene therapy. The DNA sequences are introduced into the genome to replace the defective gene fragment that has lost normal function.

Gene gun: It is a method of gene transfer mainly used in plant cells. DNA coated tungsten or gold particles are shot at a cellular structure. The energy to shoot the particles comes from an explosive charge (such as compressed gas or from an electrical field acceleration). In electrical field acceleration, the particles are launched in vacuum which means that it is not an useful method for living organisms. However, the efficiency of the method is found good in plants. Introduction of tungsten particles into humans has been questioned. Therefore, it has remained only a research tool. Gene therapists are searching more novel procedures to introduce therapeutic genes for tumor treatment. Jet stream application device has been used to inject proteins at high pressure, The same with some modifications may be considered for DNA transfer.

Gene inhibition therapy (at nucleic acid level): A novel approach where it is considered to target correction at the RNA level by using ribozymes or RNA editing enzymes to correct pathogenic mRNAs. Alternately, the approach may be used to inhibit gene expression by taking advantage of certain nucleic acids (antisense genes).

Gene inhibition therapy (at protein level): It is considered to target inhibition of gene expression at the protein level. The cell is made to express with in a cell of genetically engineered antibodies which bind to and inactivate mutant proteins. An another novel approach is to use degenerate oligonucleotides to identify specific oligonucleotide sequences which interact with proteins. These oligonucleotides or aptamers can then be used to inactivate specific mutant proteins.

Gene therapy: A technique by which DNA sequences encoding specific gene are delivered to cell with the aim of treating or curing a genetically linked disease. In genetic diseases (disorders) known to be associated with a simple incorrect nucleotide sequence in a gene, it may be possible to correct the defect using gene manipulation.

Germinal gene therapy: In this therapy, transgenic cells are introduced into the germ line as well as into the somatic cell population.

Hemophilia: is a genetic disease. The patient is unable to synthesize factor VIII or factor IX (proteins required for blood clotting). These factors are required to participate in the chain of events of blood clotting leading to the formation of long fibrous protein called as fibrin. Therefore, hemophilic patients are unable to carry blood coagulation. Minor cuts or bruises which do little harm to most people can prove fatal to hemophiliacs.

Inverse PCR: In Inverse PCR, amplification of the DNA of unknown sequences flanking a known sequence is done by using two primers facing outwards.

Isopropylthio-β-D-galactoside (IPTG): It is an inducer of β-galactosidase (*lac* Z) gene. However, it is not hydrolyzed by β-galactosidase enzyme.

Kanamycin/ Neomycin resistant gene (Kan^r): It is a gene whose product causes kanamycin resistance to the host cell. The gene codes for a protein having the molecular weight nearly 25,000. The protein gets secreted in the periplasmic space and acts as aminoglycoside phosphotransferase. It causes phosphorylation of the kanamycin as well as neomycin in the presence of ATP. Phosphorylation of the antibiotics interferes with their active transport into the cell.

Klenow fragment: Larger fragment of DNA polymerase I of *E coli.* It has only 3',5' exonuclease activity in addition to DNA polymerase activity. It lacks 5',3' exonuclease activity.

Lambda gt10 (λgt10) vector: It is derived from lambda phage. It is considered to be a non-expression vector and presence of the recombinant in it is screened with the help of clear and turbid plaques. It has only one cloning site, EcoR1 present with in the cI gene.

Lambda gt11 (λgt11) vector: It is derived from lambda phage. It is considered to be an expression vector and presence of the recombinant in it is screened with the help of blue and white plaques (histochemical screening). It has only one cloning site, EcoR1 present with in the *lac* Z gene, 53 nucleotides upstream from the termination codon.

Lesch-Nyham syndrome: a genetic disease in which the patient is unable to synthesize hypoxanthine-guanine phosphoribosyl transferase enzyme. The disease leads to a bizarre impulse for self mutilation including very severe biting of lips and fingers.

Miniprep: When plasmid DNA is isolated from a small volume (1 ml or so) of transformed bacterial cell culture, it is called miniprep. Miniprep is used for quick isolation of small amount of plasmid DNA which is about 80% pure. It is used for restriction endonuclease digestion.

M13 mp vectors: These vectors are constructed by manipulation of M13 phage by Messing and group. The vectors show the property of α-complementation. Besides, the vectors have a mutiple cloning site region, also called as polycloning site region.

M13 phage: It is a single stranded DNA containing bacteriophage. It infects *E.coli* cells through F episome. After infection, it mutiplies with in the host. After multiplication, the phage particles come out of the host without lysis of the host cell. It is neither lytic or lysogenic phage. It has been manipulated in constructing vectors by Messing and coworkers and vectors are by the name M13 mp vectors.

Multiplex-PCR: In this technique of modified PCR, co-amplification of two or more target sequences with in a single sample is carried out using an unique pair of primers for each target.

Nested PCR: It involves one PCR reaction followed by the next PCR extension which amplifies the first PCR product. There two sets of primers are used, one set for amplification of DNA sequences and another set for amplification of the first product of PCR.

Northern blotting: It is a technique used for detecting RNA molecule having a specific nucleotide sequence. First, RNA sample is electrophoresed and then blotted on the nitrocellulose membrane. On the blotted membrane, the specific sequenced RNA is detected using complementary sequence containing labelled DNA probe.

Nucleic Acid: It is a polynucleotide molecule. It may have ribosyl moiety (RNA) or deoxyribosyl moiety (DNA) in its structure.

Oligonucleotide site directed mutagenesis: It is a technique used for carrying a specific point mutation in a DNA molecule. In the technique, an oligonucleotide is used having a mismatch corresponding to the mutation being planned in the DNA molecule. In genetic engineering, it is considered very powerful technique which is used for doing known manipulation in the protein sequence.

Orcinol: It is a chemical used for detecting and estimating ribosyl moiety. In nucleic acids, it is used for estimation of RNA since ribose is a constitutent of RNA.

Origin of replication (*Ori*): It is the DNA sequence at which DNA replication is started (initiated).

pBR322: It is one of the very common bacterial plasmid constructed by Bolivar and Rodriguez. It has 4362 bp in its structure and is a relaxed plasmid having ColE1 replicon. The plasmid has tetracyclin as well as ampicillin resistant genes (Amp^r and Tet^r).

Phagmid vectors: These are the vectors constructed by combination of plasmids and phage sequences.

Plasmid: It is a bacterial replicon stably inherited in an extra chromosomal state.

Polymerase chain reaction (PCR): It is a technique used to amplify small DNA fragments a billion fold. It is based on the principle of DNA replication and use of heat stable DNA polymerase has been taken for PCR reaction. In conventional PCR, use of two primers facing each other is taken.

Polynucleotide kinase: It is an enzyme that catalyzes the transfer of γ-phosphoryl group from ATP onto the 5' –OH (dephosphorylated) ends of a DNA molecule. It may use single stranded as well as double stranded DNA. In case of double stranded DNA, it should have 5'-protruding or blunt ends. Polynucleotide kinase also catalyzes the exchange of the phosphoryl group between γ-phosphoryl group of ATP and phosphoryl group present at the 5'- ends of the DNA molecule. It is used for radio active labelling of the 5'-end of DNA for use as probe and for DNA sequencing during chemical modification method of Maxam and Gilbert.

Promoter: The region of DNA involved in the binding of RNA polymerase enzyme to initiate transcription.

pUC vectors: These are the vectors constructed by Messing and coworkers. These are derivatives of pBR322 plasmid. Here, UC denotes University of California where Messing and coworkers constructed these vectors. The average size of pUC vectors is nearly 2.7 Kb.

pUC vectors have ampicillin resistant gene (Amp^r), therefore render the host ampicillin resistant.

Recombinant vector: A vector after cloning a DNA fragment in it.

Relaxed plasmid: It is the plasmid maintained as high copy number per cell. Copy number of the plasmid is related to *rop* gene which codes for rop protein. The rop protein has negative regulation on replication.

Restriction endonuclease: It is an enzyme that binds specifically to and cleave double stranded DNA at specific site with in or adjacent to a particular sequence known as recognition sequence. In genetic engineering, endonucleases have an important place since in fact genetic engineering (specifically cloning work) became possible only due to restriction endonucleases.

Reverse transcriptase: It is an enzyme that catalyzes the synthesis of DNA using RNA as a template. It may also use DNA as a template. The enzyme has been shown to have RNase H activity which is capable of chewing RNA portion in a RNA-DNA hybrid. The enzyme has been reported from RNA viruses viz. Avian Myeloblastosis virus, Molony Murine Leukemia virus.

Reverse transcriptase polymerase chain reaction (RT-PCR): It is the modified technique of PCR which is used if RNA has to be used as template and amplification of its complementary DNA has to be carried out. During the RT-PCR, use of two enzymes namely reverse transcriptase and heat stable DNA polymerase has been taken. Nowadays, scientists have also performed RT-PCR using single heat stable reverse transcriptase since reverse transcriptase may use RNA as well as DNA as template.

Ribonucleic acid: It is a polynucleotide having ribonucleotides as its constituents.

***Rop* gene:** It codes for a protein called as rop protein which reinforces the negative regulation of replication.

Shuttle vector: If a vector can replicate in different organisms, it is called a shuttle vector. It contains components like origin of replication, promoters etc so that it may replicate in different organisms.

S1 Nuclease: It is an enzyme that catalyzes the degradation of single stranded nucleic acids to yield 5'- phosphoryl mono or oligonucleotides. Double stranded DNA, RNA and DNA-RNA hybrids are resistant to S1 nuclease. In molecular biology, it is used to remove single stranded tails from DNA fragments to produce blunt ends for ligation of specific linkers. It is also used for removal of hairpin structure during cDNA library formation.

Southern blotting: It is a technique used for detecting DNA molecule having a specific nucleotide sequence. First, DNA sample is electrophoresed and then blotted on the nitrocellulose membrane. On the blotted membrane, the specific sequenced DNA is detected using complementary sequence containing labelled probe.

Somatic gene therapy: It focuses only on the body, or soma attempting to effect reversal of the disease phenotype by treating some somatic tissues in the affected individual. In such cases, it is found that not all the cells of that tissue need to become transgenic. A portion of transgenic cells can ameliorate the overall disease symptoms. The methods proceed by removing some cells from an individual with the defective genotype and making these cells transgenic through the introduction of copies of the cloned wild type gene. The cells are then reintroduced into the patient's body, where they provide normal gene function.

Stringent plasmid: It is the plasmid maintained as low copy number per cell. Copy number of the plasmid is related to *rop* gene which codes for rop protein. The rop protein has negative regulation on replication.

Taq DNA polymerase: It is a heat stable DNA polymerase generally used in polymerase chain reaction. It has been isolated from *Thermas aquaticus*.

Terminal deoxynucleotidyl transferase: It is an enzyme that catalyzes the transfer of deoxynucleotides on to the 3' ends of DNA strands and the reaction is template independent using deoxyribonucleoside triphosphate as precurssor. The enzyme has been used in molecular biology for 3' ends labelling of the DNA molecules having 3'-protruding ends using α-P^{32}-dATP.

Tetracycline resistant gene (tet^r): It is a gene whose product causes tetracycline resistance to the host cell. The gene codes for a protein having 399 amino acids in its chain. The protein remains associated with the cell membrane and prevents the tetracycline from entering the bacterial cell.

Vector: It is a replicon stably inherited in an extra chromosomal state. Vector is a general term used for bacterial plasmids as well as phage vectors.

Western blotting: It is a technique based on the immunological principle of specific antigen antibody binding used to detect small amount of the protein. First protein sample is electrophoresed using sodium dodecyl sulfate polyacrylamide gel electrophoresis for separating proteins and thereafter, the separated proteins are blotted onto a nitrocellulose or a similar type membrane using electroblotting. Afterwards, presence of a specific antigen is detected on the blotted membrane using specific antibodies which are detected using secondary antibody conjugated either with the radio active iodine or with alkaline phosphatase.

Index

A

B

C

D

E

F

G

H

Q

R

S